The Essence of Turbulence as a Physical Phenomenon

Arkady Tsinober

The Essence of Turbulence as a Physical Phenomenon

With Emphasis on Issues of Paradigmatic Nature

Arkady Tsinober
School of Mechanical Engineering
Tel Aviv University
Tel Aviv, Israel

ISBN 978-94-007-7179-6 ISBN 978-94-007-7180-2 (eBook)
DOI 10.1007/978-94-007-7180-2
Springer Dordrecht Heidelberg New York London

Library of Congress Control Number: 2013947480

Printed on acid-free paper

Springer is part of Springer Science+Business Media (www.springer.com)

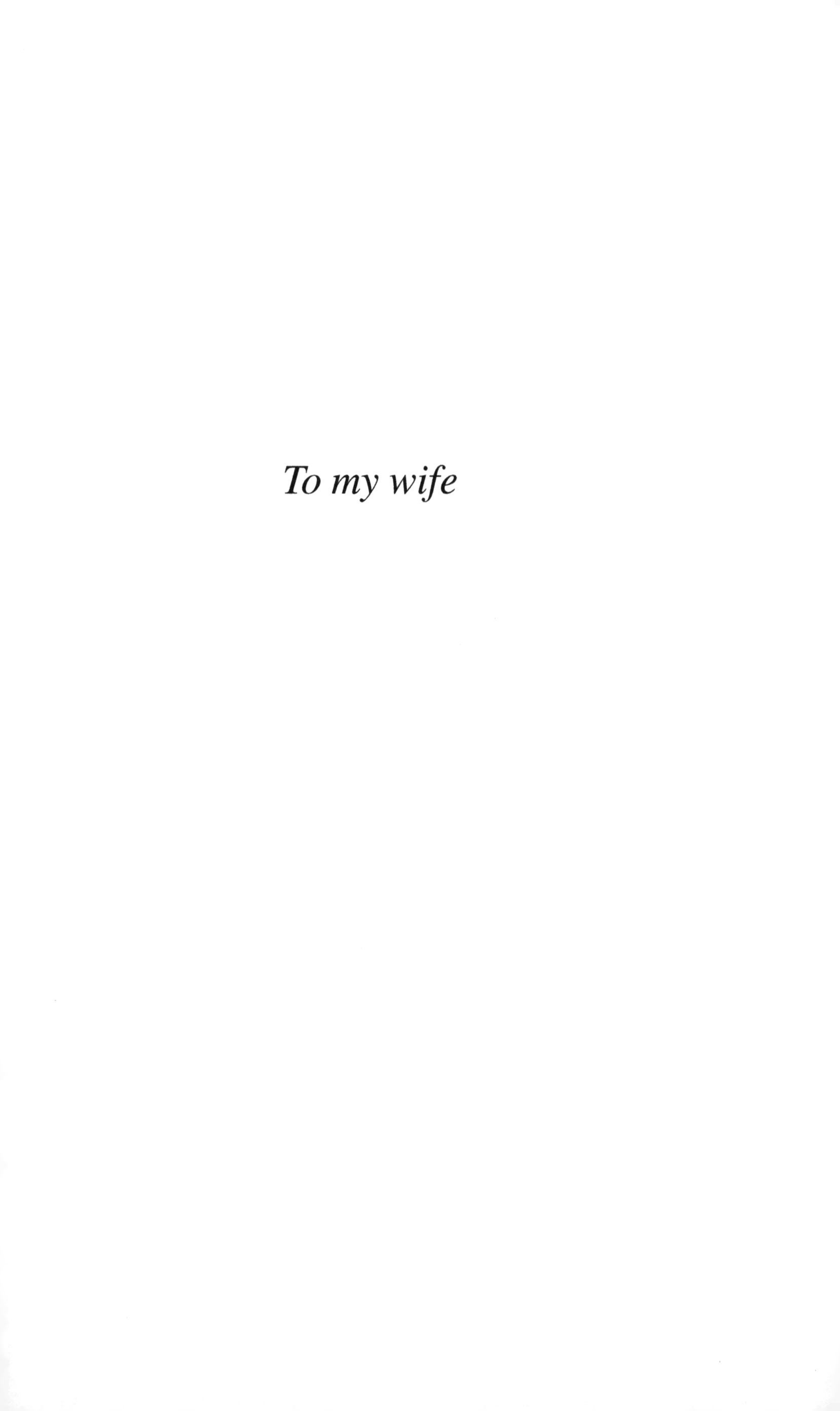

To my wife

Preface

One of my hopes is that this small book will aid the readers, especially the young ones, to develop their own judgment on the huge literature in the field and to make some distinction between the "essence" and "the water". It is also among the reasons that the main themes of this small book are about fundamental issues of paradigmatic nature. The reasons for using the term "issues of paradigmatic nature" instead of, say, just "paradigms in turbulence" is due to the question on the very existence of the latter (so far).

The main premises and the reasons/causes that it was possible to make this book short are the absence of theory based on first principles and inadequate tools to handle both the problem and the phenomenon of turbulence. Obviously, for this reason the "essence" appears implicatively rather than explicitly. This state of matters was and is continuously stressed in the community including many prominent scientists and among them by Batchelor (1962), Kolmogorov (1985), Liepmann (1979), Lumley and Yaglom (2001), von Neumann (1949), Ruelle (1990), Saffman (1978, 1991), Wiener (1938), see Appendix for essential quotations.

Indeed, the heaviest and the most ambitious armory from theoretical physics and mathematics was tried for more than fifty years, but without much success: genuine turbulence, the big T-problem, as a physical and mathematical problem remains unsolved.

It is not a trivial task to address things like paradigmatic issues (as, in fact, there are no real paradigms in fundamental research of turbulence) in a field which—in words of Liepmann is the *graveyard of theories*, so that one has to relay mainly on the empirical evidence. The task is not made easier in view of no consensus on what is (are) the problem(s) of turbulence, neither is there an agreement on what are/should be the aims/goals of turbulence research/theories and what would constitute its (their) solution.

The basic properties of turbulence of fundamental nature and *the essential mathematical complications of the subject were only disclosed by actual experience with the physical* (and numerical) *counterparts of these equations* (von Neumann 1949) and of the theoretical objects in question many of which still await to be found and properly defined. Today, as before, observations remain the major exploratory tool in elucidating the properties of turbulence as a physical phenomenon. This is not to claim absence of theory(ies). On the contrary, there are plenty—many with qualitatively different and even contradictory premises—all agreeing well with *some* experimental data and even claiming rigor, but not necessarily for the right reasons and not based on first principles with few exceptions having no direct bearing on the Navier–Stokes equations and thereby being mostly out of scope of this book.

Paradoxically, due to the alarming overabundance and continuing major overproduction of publications of theoretical nature claiming "theory" or exploiting the term *turbulence* in one way or another, the above task is made easier. This is because most of such publications did not bring any real breakthrough in understanding and have not much to do with the essence of turbulence without contributing to alleviating the long-lasting and continuing paradigmatic crisis in the basic research of turbulence. There was still a need for the "Saffman's criterion" (Saffman 1978), in order to reduce the amount of references but complemented by typically giving some early and latest references. The book by the author (Tsinober 2009), contains more than one thousand references on basic issues, so that it may be of help also in the context of this small book as the former is devoted also exclusively to the basic issues.

Still in such a situation one has to make difficult choices on what one has to put on paper. It is obvious, however, that the choice has to be made out of issues of fundamental/basic nature which are based on fundamental observations/facts and closely related issues.

The book is limited by incompressible flows and consists of three parts and an Epilogue. Part I is devoted to main aspects of what is both the nature of the phenomenon and the problem of turbulence, what equations describe turbulence adequately, and also inadequately, along with some closely related issues. Part II is about brief description of what are the origins of turbulence and in a bit more detail about the nature of both the phenomenon and the problem with the emphasis on discussion of various facets of the "undeniably statistical nature" contrasted to the "deterministic origin" of turbulence as described by purely deterministic non-integrable equations and related issues. This includes a discussion of the consequences of complex behavior of systems described by purely deterministic equations such as the necessity of change of the paradigmatic meaning of apparent randomness, stochasticity of turbulence which is roughly just the complexity due to a large number of strongly interacting degrees of freedom governed by the Navier Stokes equations. Part III is a continuation of the previous part, and covers more specific issues, but not less important. What follows comprises or are related to the second part of the major qualitative universal features of turbulent flows briefly described in Chap. 1. A distinction is also made between the issues of paradigmatic nature and those which are apparently/seemingly, i.e. pseudo-paradigmatic such as the most popular cascades.

The Epilogue is a brief reiteration of the main points with different more general accents and reminding that all ideas/theories, etc. proposed so far did fall/failed belong to the category of misconceptions and/or ill defined concepts, see Chap. 9 in Tsinober (2009). These are followed by remarks on what next.

Tel Aviv, Israel A. Tsinober
March, 2013

Contents

Part I
The Phenomenon and the Problem of Turbulence

Let us first understand the facts, and then we may seek for their causes (Aristotle 384–322 BC).

Remember, when discoursing about water, to induce first experience, then reason (Leonardo da Vinci 1452–1519).

We are certainly not to relinquish the evidence of experiments for the sake of dreams and vain fictions of our own devising (Isaac Newton, Preface to his Principia Mathematica, Motte translation as revised by F. Cajori, Berkeley, 1947, pp. xvii–xviii).

Studying nature, which is what natural sciences do, requires a flexible approach based on facts and not on dogmas (Ginzburg 2003, Autobiography).

It is of importance to make at the outset a clear distinction between the phenomenon and the problem of turbulence—just like between the observations and the theories attempting to explain them, which are far from being synonymous. One has also to make a distinction between the nature of the phenomenon and of the problem, for example, one cannot qualify the phenomenon of turbulence just as a purely statistical object. The statistical/probabilistic nature is one among numerous attempts to put turbulence into the Procrustean bed of some specific field pretty arbitrarily and essentially just because it is more than very complex so that we are still quite disoriented as to the relevant factors, and as to the proper analytical machinery to be used (von Neumann in Collected works, vol. 6, pp. 437–472, 1949). Statistics is an approach in the first place, so one can hardly speak about statistical, probabilistic, etc. nature of turbulence as a phenomenon. In contrast, one can easily assign a statistical/random as contrasted to deterministic "nature" to a theory—with the question whether such a theory is adequate to the phenomenon of turbulence. In other words, theory—if such exists—is the human representation of the phenomenon and it is important to make a distinction with the phenomenon itself.

Indeed, unlike other complicated phenomena it is easy to observe at least some of the numerous manifestations of turbulence as a physical phenomenon, but it is extremely difficult to comprehend, interpret, understand and explain—all this comprising the problem of turbulence.

The rest of this Part I is devoted to which equations describe turbulence adequately, and also inadequately, along with some closely related issues.

Chapter 1
The Phenomenon of Turbulence as Distinct from the Problem of Turbulence

Abstract The issue is about turbulence as a natural physical phenomenon as related to observations as distinct from the problem of turbulence. The dichotomic distinction between laminar and turbulent flows is problematic in several respects. First, most flows termed "turbulent" are in reality partly turbulent: some portions of the flow as turbulent and some as laminar—the coexistence of the two regimes in one flow is a common feature with continuous transition of laminar into turbulent state via the entrainment process through the boundary between the two. Moreover, the reality is not that simple as the laminar/turbulent dichotomy as, e.g. the behavior of passive objects in flows with small Reynolds number looks as perfectly turbulent reflecting the qualitative difference between the chaotic flow properties in Eulerian and Lagrangian settings. These examples illustrate the enormous difficulties in defining what is both (i) turbulence and (ii) the turbulence problem. As concerns (i) one can only provide a description of major qualitative universal (sic) features of turbulent flows as obtained almost exclusively from observations (rather than by any theoretical deliberations) which form most important part of the "essence" of turbulence. This is because these mostly widely known qualitative features of all turbulent flows are essentially the same, i.e., it is meaningful to speak about qualitative universality of turbulent flows. It has to be stressed that the term "phenomenon of turbulence" as used above is mostly associated with the observational aspects, which in turbulence play far more important role due the unsatisfactory state of "theory": there seems to exist no such a thing based on first principles. Hence it is vital to put the emphasis on the physical aspects based in the first place on observations.

The issue is about turbulence as a natural physical phenomenon as related to observations both passive and active.

A typical starting statement on turbulence is as follows:

It is known that all flows of liquids and gases may be divided into two sharply different types; the quiet smooth flows known as "laminar" flows, and their opposite, "turbulent" flows in which the velocity, pressure, temperature and other fluid mechanical quantities fluctuate in a disordered manner with extremely sharp and irregular space- and time-variations (Monin and Yaglom 1971, p. 1).

A. Tsinober, *The Essence of Turbulence as a Physical Phenomenon*,
DOI 10.1007/978-94-007-7180-2_1,

Fig. 1.1 Side view of a turbulent boundary layer visualized by smoke traces, courtesy of Professor H. Nagib

However, as common as it is, this division is problematic in several respects. First, looking at Fig. 1.1 one can easily see that the flow is only partly turbulent (PTF).

Most flows termed "turbulent" are of this kind: boundary layers, all free shear turbulent flows (jets, plumes, wakes, mixing layers), penetrative convection in the atmosphere and in the ocean, gravity currents, avalanches and other phenomena at the boundary between single phase fluid and fluid loaded by a sediment (which includes resuspension), clear air turbulence, and many others (e.g., combustion). Transitional flows consisting as a rule of turbulent regions growing in a laminar environment (such as turbulent spots) are also partly-turbulent flows.

However, it is does not look as conceptually correct to say that a turbulent flow (not a partly turbulent flow) is completely laminar locally, e.g. at small scales with locally small Reynolds number, as in the examples belonging to a span of time period of 65 years (Batchelor 1947; Southerland et al. 1994; Hamlington et al. 2012 and references therein). Another example is a problematic statement that "the large scale circulation of the atmosphere is non-turbulent" though "the smaller scale flow is certainly turbulent" (Cullen 2006).

In these flows one observes some portions of the flow as turbulent and some as laminar—the coexistence of the two regimes in one flow is a common feature with continuous transition of laminar into turbulent state via the entrainment process through the boundary between the two. All these flows are also neither homogeneous nor isotropic with some other "N's" , which has implications of using an interpretation the data for comparisons, "validation" of theoretical stuff based on homogeneous, isotropic and/or flows with periodic BC's.

Both PTF's and "purely" turbulent flows exhibiting a high degree of apparent randomness and disorder often reveal the presence of what is termed frequently as ordered embedded flow structures.

Another worthy mentioning is a class of very intense objects such as rotating storms in the atmosphere again existing in and coexisting with "ordinary"

Fig. 1.2 May 28, 2004 Highway 12 Nebraska Supercell. http://www.spc.ncep.noaa.gov/exper/archive/event.php?date=20040528

flows/regimes which, see Fig. 1.2. The intense objects are generally strongly turbulent whereas the ordinary environment may be even close to a laminar state.

Moreover, the reality is not that simple as the laminar/turbulent dichotomy as illustrated by Fig. 1.3 in which the flow looks perfectly turbulent. In fact, this is a flow at Reynolds number of order unity with an essentially linear dependence between the flow rate and the pressure drop and other dynamic attributes of laminar flows at low Reynolds numbers. There is a great variety of flows with Reynolds number effectively close to zero, i.e. laminar in Euler setting (E-laminar) but is chaotic in Lagrangian setting (L-turbulent), for references (Tsinober 2009).

This is because what looks to be turbulent by employing visualizations such as via observing passive objects, e.g. using some dye, may have nothing to do with the behavior of dynamic variables such as velocity and vorticity and even with the turbulent nature of the flow in question, and reflecting the chaotic nature of the behavior of passive objects in a purely laminar flow in an Eulerian setting called also Lagrangian/kinematic chaos, or "Lagrangian turbulence" as distinct from the chaos in Eulerian setting. Another example is about the "streaks" in a turbulent flow in a plane channel (Baig and Chernyshenko 2005). Although the vortical structure of the flow is the same, the scalar streak spacing varies by an order of magnitude depending on the mean profile of the scalar concentration. Moreover, passive-scalar streaks were observed even in an artificial "structureless" flow field.

The above examples illustrate the enormous difficulties (and impossibility, so far) in defining what is both (i) turbulence/turbulent motion[1] and (ii) the turbulence problem. As concerns (i) one can only provide a description of major qualitative

---

[1]Though there are many attempts to do so. In contrast to mathematical *theories* in which the definition of the main object of the theory precedes the results, in turbulence (as in any field of physics) even if such a definition would be possible it is likely to come after the basic mechanisms of turbulence as a physical phenomenon are well understood.

In any case there is considerable 'turbulence' in the attempts to define what is turbulence indicating that such attempts at the present stage are futile not to mention that no adequate theory is in existence.

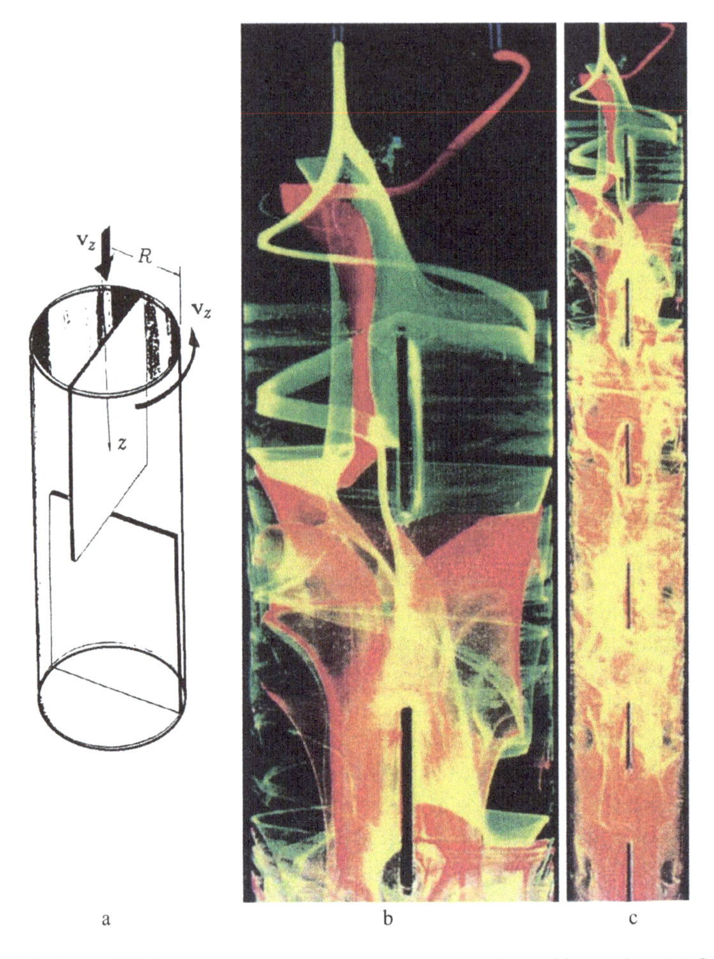

Fig. 1.3 Mixing in PPM—partitioned-pipe mixer at very low Reynolds number. (**a**) Schematic of the PPM, (**b**) is a close-up of the upper part of (**c**). From Kush and Ottino (1992). For other examples see references in Tsinober (2009)

universal (sic) features of turbulent flows as obtained almost exclusively from observations rather than by any theoretical deliberations, see the quotation by von Neumann (1949) in the Appendix essential quotations.

1.1 Major Qualitative Universal Features of Turbulent Flows

The list below contains a brief description of the major qualitative universal features of turbulent flow. Important quantitative and more specific features will be discussed in more detail in the sequel. These altogether form most important part of the "essence" of turbulence.

- Intrinsic spatio-temporal apparent randomness. Turbulence is definitely chaos. However, vice versa, generally, is not true: many chaotic flow regimes are not necessarily turbulent, e.g. Lagrangian/kinematic chaos or 'Lagrangian turbulence' (L-turbulent), but laminar in Euler setting flows (E-laminar). One of the most important aspects is that the apparent stochastic/random nature of turbulent flows is its intrinsic property,[2] i.e. self-stochastization or self-randomization. We emphasize the term *apparent* randomness/stochasticity: after all turbulent flows are known (so far) to be governed by the NSE which are purely deterministic, so that there is quite a bit of a problem with the random/deterministic dichotomy either just like with the use of term randomness and its synonyms.
- Extremely wide range of strongly and non-locally interacting degrees of freedom which are not necessarily synonymous to the most popular modes of some decomposition. Turbulent flows are very large systems. In atmospheric flows, relevant scales range from hundreds of kms to parts of a mm, i.e., there exist $\sim 10^{29}$ excited degrees of freedom, many of which are strongly interacting. Hence extreme complexity of turbulence enforcing the statistical description of turbulent flows. We emphasize again that statistical description is not synonymous to statistical theorization not to mention statistical "nature".
- Loss of predictability, which is one of the attributes of the chaotic nature of turbulence. Two initially nearly (but not precisely) identical turbulent flows become unrecognizably different on the time scale of dynamic interest. However, generally, different realizations of the same turbulent flow have the same statistical properties. In this sense the statistical properties (not only some means, but almost all statistical properties) of turbulent flows are insensitive to disturbances—turbulent flows are statistically stable as possessing statistically stable properties.

  The above items are discussed in Part II, the rest in Part III.
- Turbulent flows are three-dimensional, highly dissipative and thereby time irreversible and rotational, i.e., carry lots of strain and vorticity. Thus the field of velocity derivatives is of special importance. One of the key physical processes is the predominant production of the velocity derivatives, to stress both strain/dissipation and vorticity as almost equal partners. "Almost" is because it is the strain production is responsible for production of both contrary to common view about amplification vorticity. These attributes are, probably, the most specific and important as concerns fluid dynamic turbulence. On the paradigmatic level it is the nonlinearity that is responsible for these most basic key properties of turbulence as essentially rotational and strongly dissipative phenomenon. It is noteworthy that the predominant production of velocity derivatives is irreversible, i.e. not just dissipation is responsible for the irreversibility.

  There is no consensus whether two-dimensional chaotic flows even with many degrees of freedom should be qualified as turbulence. The main reason is that

[2] Cf. with the case of random boundary/initial conditions and random noise on the RHS of NSE. Of special interest for this comparison are systems/equations which do not exhibit any quasi-random behavior without external stochastic excitation. For example, Burgers and Korteveg de Vries equations or just the NSE at small Reynolds numbers.

such flows lack the mechanism of vorticity and strain amplification and weak excitation of small scales.

- Strongly diffusive (random waves are not). Turbulent flows exhibit strongly enhanced transport processes of momentum, energy, passive objects (scalars, e.g., heat, salt, moisture, particles; vectors, e.g., material lines, gradients of passive scalars, magnetic field). It should be emphasized that in respect with passive objects this property is true of a much broader class of systems. Namely, any random velocity field and even laminar flows in Euler setting, which are Lagrangian chaotic, exhibit enhanced transport of passive objects.
- Strongly nonlinear, non-integrable, nonlocal, non-Gaussian with some more N's, see Chap. 7.

These mostly widely known qualitative features of all turbulent flows are essentially the same, i.e., it is meaningful to speak about qualitative universality of turbulent flows. We discuss this issue in Chap. 8.

It has to be stressed that the term "phenomenon of turbulence" as used above is mostly associated with the observational aspects, which in turbulence play far more important role due the unsatisfactory state of "theory": there seems to exist no such a thing based on first principles. It is vital to put the emphasis on the physical aspects based in the first place on observations with the distinction between "active" and "passive". Otherwise, such an endeavor becomes pretty problematic due to the highly dimensional nature of the phenomenon: one can do a continuum of things leading to nowhere or at best with very small useful output. In other words, we stress that the observational aspect is not that trivial in such a highly-dimensional system as turbulence. It is intimately related to the skill/art of asking the right and correctly-posed questions: one has to have at least some idea what to look at and for what reason, i.e. one has to have some idea about what is (are) the problem(s) of turbulence. The latter brings us to the issue of the problem of turbulence.

Chapter 2
The Problem of Turbulence as Distinct from the Phenomenon of Turbulence

Abstract The issue involves a set of questions concerning theory(ies) of turbulence and questions such as what is physics and what is mathematics of turbulence and what are the physical/mathematical problems of turbulence.

Though there is no acceptable definition of (what is) turbulence as a physical phenomenon, the relative "easiness" of observing its diverse manifestations leaves less problems than in the above. There is no consensus on what is (are) the problem(s) of turbulence and what would constitute its (their) solution. There are even doubts about the very existence of the problem or any essence in it, so that the failure so far of theoretical efforts may be blamed on the problem itself. Neither is there agreement on what constitutes understanding. There is also no consensus on what are the main difficulties and why turbulence is so impossibly difficult: almost every aspect of turbulence is controversial, which by itself is one of the greatest difficulties. As concerns the basic aspects of the problem there is far more to say about the difficulties rather than achievements. The problem was recognized by Neumann and Kolmogorov among others.

The most acute difficulty is of basic and conceptual nature and concerns the physics of turbulence and the lack of knowledge about the basic physical processes of turbulence, its generation and origin, and poor understanding of the very few processes which are already known. One of the key physical processes is the predominant production of the velocity derivatives, to stress both strain/dissipation and vorticity as almost equal partners. "Almost" is important because it is the strain production is responsible for production of both contrary to common view about amplification of vorticity. These attributes are, probably, the most specific and important as concerns fluid dynamic turbulence. Another set of much neglected issues is about the nonlocal properties of turbulence and related questions such as the ill posedness of the concepts/paradigms of inertial range and cascade, and the role of large scales and viscosity/dissipation; scale invariance, symmetries and universality in turbulence; origins of intermittency and the so-called 'anomalous scaling in the inertial range' just to mention some.

This state of matters to a large extent can be qualified as long-lasting and continuing paradigmatic crisis.

A. Tsinober, *The Essence of Turbulence as a Physical Phenomenon*,
DOI 10.1007/978-94-007-7180-2_2,

The issue involves a set of questions concerning theory(ies) of turbulence and questions such as what is physics and what is mathematics of turbulence and what is the physical/mathematical problem of turbulence; what equations describe turbulence adequately. The very question is a problem—how one can decide whether, say, NSE are adequate or what are the equations of turbulent motion, see Chap. 3.

Though there is no acceptable definition of (what is) turbulence as a physical phenomenon, the relative easiness of observing its diverse manifestations leaves less problems than in the question what is/comprises (are) the scientific, mathematical and physical, problem(s) of turbulence. There is no consensus on what is (are) the problem(s) of turbulence and what would constitute its (their) solution.[1] Neither is there agreement on what constitutes understanding. Also just like no sophisticated experiment, laboratory or DNS,[2] by itself does not bring understanding, neither does modeling of whatever sophistication. This can be brought only by a genuine theory, which seems to be not in existence so far. Though there exists a set of deterministic differential equations, the Navier–Stokes equations (NSE), probably containing (almost) all of turbulence, most of our knowledge about turbulence comes from observations and experiments, laboratory, field and later numerical, which is unfortunate as theory is supposed to guide and gives meaning to observation.[3] This was under-

[1]There are even doubt about the very existence of the problem or any essence in it, so that he failure so far of theoretical efforts may be blamed on the problem itself:

I hope that my talk will help to clarify a number of the problems and I think that some of the models of chaotic phenomena that I will discuss are illuminating and suggestive, but it is far too early to claim that any of them gives the essence of the phenomenons of turbulence—if indeed it is a single phenomenon and has an ***essence*** (Martin 1976).

...we should not altogether neglect the possibility that there is no such thing as 'turbulence'. That is to say, it is not meaningful to talk of the properties of a turbulent flow independently of the physical situation in which it arises. In searching for a theory of turbulence, perhaps we are looking for a ***chimera*** (Saffman 1978).

However, today there is no doubt about the existence of a set of qualitatively universal properties comprising the essence of any three-dimensional turbulent flow. This is result of dealing in the first place with the phenomenon itself, rather than with "theories", though excessively and quite ineffectively.

[2]*Progress in numerical calculation brings not only great good but also awkward questions about the role of the human mind... The problem of formulating rules and extracting ideas from vast masses of computational or experimental results remains a matter for our brains, our minds* (Zeldovich 1979).

There are essential differences between physical and numerical experiments. If underresolved the former still provide correct information which is a problem with the latter especially as concerns numerical errors sometimes interpreted as genuine chaos.

[3]This is the main reason that this book is biased experimentally and this is why the importance of experimental research in turbulence goes far beyond the view of those who think of an experimentalist as a superior kind of professional fixer, knowing how to turn nuts and bolts into a confirmation of their theories. From the basic point of view there is almost nothing to be confirmed so far. In absence of true theory and in view of the high dimensional nature of the problem the issue of confirmation/validation of 'theories' in turbulence is far more serious than just checking the 'validity' of some formulae or any other theoretical stuff via comparison with limited experimental evidence, especially because turbulence is a highly dimensional problem.

stood long ago by many outstanding people, Kolmogorov, Von Neumann, Wiener and many other, see references and quotations for the preface in the Appendix.

Indeed, the basic properties of turbulence of fundamental nature such as the major features listed above among others *were only disclosed by actual experience with the physical* (and numerical) *counterparts* of the theoretical objects in question many of which still await to be found and properly defined. Today, as before, the experiment (physical and numerical) remains the major exploratory tool in elucidating the properties of turbulence as a physical phenomenon. This is not to claim absence of theory(ies). On the contrary, there are plenty—many with qualitatively different and even contradictory premises/assumptions—all agreeing well with **some** experimental data, but not based on first principles, i.e. with few exceptions have no direct bearing on the Navier–Stokes equations and thereby being mostly out of scope of this book as lacking the main attribute of a true theory:

The problem of turbulence is not just to find more accurate formulae for various physical quantities associated with a turbulent fluid, but also to obtain a conceptually satisfactory theory based on first principles... In spite of satisfaction which one may have in writing rigorous inequalities originating from nontrivial linear theory, it must be said that the great difficulty which remains is to understand the nonlinear objects of turbulence (Ruelle 1990). But as mentioned 'understanding' is a vague concept, just like it is far from being clear enough what are 'the nonlinear objects of turbulence', though some would jump on 'coherent structures' which along with some other objects called "structures" still are pretty elusive and share quite a bit in common with the Emperors's new Clothes. Unfortunately, this is true of several other aspects of turbulent research.

In general terms and in absence of true theory the physical/mathematical problem is to identify, understand/interpret and explain the major basic/fundamental physical mechanisms that, e.g. result in the major qualitative universal properties of turbulent flows as described above along with some other specific fundamental issues. This includes the identification and study of the significant nonlinear objects of turbulence, i.e. what objects are right to look at and requires non-trivial efforts from theoretical physicists and mathematicians. Reiterating, so far, neither theoreticians nor the mathematicians were able to achieve any of these goals.

Indeed, though the heaviest and the most ambitious armory from theoretical physics and mathematics was tried for more than fifty years, as concerns the basic/fundamental aspects, the progress is far less than modest and the state of matters is not much different from that described by Batchelor (1962), Liepmann (1979), von Neumann (1949), Saffman (1968, 1991) along with many other famous scientists, see Appendix essential quotations and Tsinober (2009).

Along with widely recognition that turbulence is formidably difficult,[4] just like there is no consensus on what is (are) the problem(s) of turbulence and what would

[4]The mood was not always that skeptical. Here are two examples from outstanding people.

Hopf (1952): *My attempts at finding the "relevant" solutions of the α-equation have been unsuccessful even in the simplest case of boundary-free flow, but I believe that the mathematical difficulties of the problem arise from the fact that it is unprecedented and not from any intrinsic complexity.*

constitute its (their) solution, there is no consensus on what are the main difficulties and why turbulence is so impossibly difficult: almost every aspect of turbulence is controversial, which by itself is one of the greatest difficulties. As concerns the basic aspects of the problem there is far more to say about the difficulties rather than achievements. The difficulties mentioned by Neumann and Kolmogorov among others comprise only a part of the whole set of difficulties. Naturally, this is one of the themes throughout this small book along with the emphasis on the key issues of turbulence as a physical phenomenon.

2.1 On Physics and Mathematics of Turbulence

One of the key words is—as expected—difficulties, so that a considerable place is taken here discussing this issue.

The most acute difficulty is of basic and conceptual nature and concerns the physics of turbulence and the lack of knowledge about the basic physical processes of turbulence, its generation and origin, and poor understanding of the very few processes which are already known. For example, the underlying mechanisms of predominant vortex stretching, which is why in turbulent flows vorticity is stretched more than compressed, are poorly understood and essentially not known. Until recently a not less important concomitant process of strain production was mostly neglected by the community. It is this process, rather than vortex stretching, that is directly responsible for the enhanced dissipation of turbulent flows and even feeds the process of predominant vortex stretching as well. There are qualitative differences between the two. The enstrophy production is a nonlocal process, whereas the strain production is mostly a local process, i.e. self-production. It is noteworthy that this "self" is conditional in the sense that the field of strain is efficient in the above two missions only with the aid of vorticity, i.e. only if the flow is rotational. Another set of much neglected issues is about a set of nonlocal properties of turbulence and related questions such as the ill posedness of the concepts/paradigms of inertial range and cascade, and the role of large scales and viscosity/dissipation; scale invariance, symmetries and universality in turbulence; origins of intermittency and the so-called 'anomalous scaling in the inertial range' just to mention some.

This state of matters to a large extent can be qualified as long-lasting and continuing paradigmatic crisis. Indeed, the basic properties of turbulence of fundamental nature, such as the major features listed above among others, were only disclosed

---

Kraichnan (1961): *In order to keep the formalism as simple as possible, we shall, work here with the one-dimensional scalar analog to the Navier–Stokes equation proposed by Burgers. In the method to be presented here, the true problem is replaced by models that lead, without approximation, to closed equations for correlation functions and averaged Green's functions... The treatment of Navier–Stokes equation for an incompressible fluid, which we shall discuss briefly, does not differ in essentials.*

The temptations with such and similar analogies are still going on.

by actual experience with the physical and numerical counterparts of the theoretical objects in question (von Neumann 1949) many of which still await to be found and properly defined. Today, as before, the observations and experiments, physical and numerical, remain the major exploratory tool in elucidating the properties of turbulence as a physical phenomenon.

Another kind/set of difficulties is mostly of a formal/technical nature and concerns the lack of adequate tools to handle the problem(s)—both theoretical/mathematical and experimental, especially at large Reynolds numbers not accessible via DNS.

Whatever the tools applied the emphasis in the sequel is made among other things on the difficulties of general nature in the contexts of conceptual and fundamental nature.

Chapter 3
What Equations Describe Turbulence Adequately?

Abstract As for today the standpoint of continuum mechanics reflected by the Navier–Stokes equations as a coarse graining over the molecular effects is considered as adequate the and "Perhaps the biggest fallacy about turbulence is that it can be reliably described (statistically) by a system of equations which is far easier to solve than the full time-dependent three-dimensional Navier–Stokes equations" (Bradshaw in Exp. Fluids 16:203–216, 1994).

One of the hot issues is on large Re, zero viscosity limit and relevance of Euler equations with the natural but futile tendency to "circumvent" in a variety of ways (as described in this chapter) the singular nature of the zero viscosity limit. A variety of empirical observations show clearly that turbulence is not a slightly viscous/dissipative phenomenon and that inertial and dissipative effects are of the same order at whatever Reynolds number and are not limited each by its own range of scales. This is manifested not only by the finite energy dissipation rate, but among other evidence, for example, by the so called Tennekes and Lumley balance between the enstrophy production (and similarly strain) and its viscous destruction/'dissipation', so that vorticity is not frozen in the flow field—again at any, however, large Reynolds numbers.

Special mentioning deserve the issue of utility of various decompositions. Being useful in many respects in the analysis of both flow states and processes decompositions have a rich potential (i) to be misused leading to misconceptions and ambiguities and introducing non-trivial artifacts and spurious effects due to properties of the decompositions not characteristic of genuine turbulence and not found in physical space and (ii) become an object of study by themselves keeping many people busy to a large extent with the properties of decompositions themselves rather than with genuine physics of turbulence. The above is among the reasons why decompositions became to some extent less productive/useful and even obscuring the physics of turbulence.

An important issue concerns the formal equivalence of Eulerian and Lagrangian representations. However, the latter appear to be 'more chaotic' with the multitude of E-laminar flows having no counterpart to the corresponding L-turbulent statistics in the same fluid flow.

One more problem is with use of functional analysis as "suspect" to belong to excessive generalizations or 'overgeneralizations' dealing with generally singular objects of unknown relevance to the physical phenomenon of turbulence. This ten-

A. Tsinober, *The Essence of Turbulence as a Physical Phenomenon*,
DOI 10.1007/978-94-007-7180-2_3,

dency for overgeneralization in science was criticized by Poincare at the beginning of the 20-th century 1902.

The title contains already a statement that some equations do which is definitely a luck. The existing evidence and the consensus is that these are the Navier–Stokes equations. This short statement has quite a bit behind. At this stage the most important is that the observed properties of turbulence statistical and not, for a given flow geometry and a given Reynolds number, are the same in air as in water and many other Newtonian fluids very different at a molecular level and share very little except the Navier–Stokes equations.

3.1 Navier–Stokes Equations

> *Perhaps the biggest fallacy about turbulence is that it can be reliably described (statistically) by a system of equations which is far easier to solve than the full time-dependent three-dimensional Navier–Stokes equations* (Bradshaw 1994).

Though the NSE have a limited kinetic foundation, they are commonly believed to be adequate in the sense that their solutions correspond to real fluid flows and, indeed, there exists large empirical evidence that NSE are adequate, at least, at all accessible Reynolds numbers. Theoretically this is not obvious, since the NSE are a gradient expansion. So in principle, higher order terms may become dominant in regions with large velocity gradients, but so far there no evidence for this.[1]

Thus as for today the standpoint of continuum mechanics reflected by the NSE as a coarse graining over the molecular effects is considered as adequate. One of the basic reasons for the success of the NSE is the existence of a gap between the scales of molecular motions and the scales of the smallest relevant scales in fluid flows including turbulence.

Thus at least formally, as concerns 'theory' the first step is to solve the Navier–Stokes equations subject to initial and boundary conditions and forcing. At present, it is possible to obtain fully resolved solutions at modest Reynolds numbers via direct numerical simulations of the Navier–Stokes equations. Such solutions are in

---

[1]But see Goldstein (1972) concerning the success of NSE equations for the laminar flows of viscous fluids, but even in this case, *it is, in fact, surprising that the assumption of linearity in the relation between* τ_{ij} *and* s_{ij} *as usually employed in continuum theory,... works as well, and over as large a range, as it does. Unless we are prepared simply to accept this gratefully, without further curiosity, it seems clear that a deeper explanation must be sought.*

Also, Ladyzhenskaya (1975) and McComb (1990), Friedlander and Pavlović (2004) on alternatives to NSE, and Tsinober (1998b, 2009) and references therein. In any case it is safe to keep in mind that no equations are Nature.

Note also the statement by Ladyzhenskaya (1969): *...it is hardly possible to explain the transition from laminar to turbulent flows within the framework of the classical Navier–Stokes theory.*

Finally, since Leray (1934) one was not sure about the (theoretical, but not observational) possibility that turbulence is a manifestation of breakdown of the Navier–Stokes equations.

some sense of the same nature as the data available from observations, but both are only necessary for the beginning of the process called understanding. However, strong arguments were given that looking at the behavior of a particular solution may not necessarily contribute much to the understanding of the basic physics of turbulent flows, so that *nothing less than a thorough understanding of the* [global behavior of the] *system of all their* [NSE] *solutions would seem to be adequate to elucidate the phenomenon of turbulence... There is probably no such thing as a most favored or most relevant, turbulent solution. Instead, the turbulent solutions represent an ensemble of statistical properties, which they share, and which alone constitute the essential and physically reproducible traits of turbulence*[2] (von Neumann 1949). That is in order to understand the dynamics, or the main characteristics of the dynamics, it is necessary to understand a significant portion of the phase flow. However, at present (if ever) it is impossible due to very high dimension and complex behavior of turbulent flows and structure of the underlying attractors (assumed to exist): *one may never be able to realistically determine the fine-scale structure and dynamic details of attractors of even moderate dimension... The theoretical tools that characterize attractors of moderate or large dimensions in terms of the modest amounts of information gleaned from trajectories* [i.e., particular solutions]... *do not exist... they are more likely to be probabilistic than geometric in nature* (Guckenheimer 1986). Note the accent on "probabilistic tools" rather than "nature" in contrast to, e.g. Foiaş et al. (2001).

In other words, one has so far to resort to statistical methods of data processing and analysis (not synonymous to statistical theorization) as the necessity imposed by extremely intricate, complex, effectively/apparently and seemingly random behavior along with a huge number of strongly and nonlocally interacting degrees of freedom.

With all this we will take the position that the basic block of the operational truth is provided by a solution of a "master" problem such as an initial, boundary conditions problem with deterministic (rather than stochastic/random) forcing for the NSE without stratification, rotation, combustion, etc. if not stated otherwise, assuming that such a problem is well posed, which seems to be the case for the NSE. Though the latter is still not rigorously proven there is evidence and consensus is that NSE most probably contain all of incompressible turbulence. We return to the issue of the dichotomy deterministic—statistical and related in Chap. 5.

There are serious issues about adequate boundary and initial conditions relevance of the Euler equation and other equations "replacing" NSE such as employing hyperviscosity, Galerkin truncations and decimated approaches among others.

---

[2] However, a far less-trivial issue is ergodicity, i.e. if the flow is statistically stationary, the common practice is to use one long enough realization, i.e. it may suffice to have such a realization at least for those who believe that statistics is enough. The basis of this is the ergodicity hypothesis, see Chap. 6.

3.1.1 On Boundary Conditions

Using the NSE for simulating real turbulent flows is quite a bit more complicated due to the problems with the boundary conditions. Among others these include inflow and outflow conditions, and problems with the so called and popular periodic boundary conditions believed to be appropriate in DNS computations to mimic (but seems that not more than that!) some artificial/idealistic flows such as statistically homogeneous turbulent flows or flows with a homogeneous direction(s). An example of a problem of conceptual nature is represented by good agreement between the results of typical DNS computations of NSE of turbulent flows (e.g., in a circular pipe and a plane channel, in a cubic box, etc.) with periodic boundary conditions with the results obtained for real flows in laboratory, in which the boundary conditions have nothing to do with periodicity and pretty frequently even not well known, and in which the initial conditions were totally different from those used in DNS. Other examples concern the role "remaining" for boundary conditions at zero viscosity limit, relevance of the Euler equation, etc.

3.2 Large Re, Zero Viscosity Limit and Relevance of Euler Equations

One of the 'natural' conjectures in the mathematical community was that turbulent flows may be described asymptotically correctly by some sort of specially selected weak, i.e. distributional solutions of the Euler equations, though these singular solutions have little to do with real physics at any however large Reynolds number. This was prompted mainly by two inputs. The first is the hypothesis on the existence at high Reynolds numbers of the so called inertial range in which viscosity/dissipation play no role (Kolmogorov 1941a), the second is the conjecture by Onsager (1949) that not smooth enough solutions of the Euler equations do not conserve energy, i.e. in some sense are "dissipative" and thereby time irreversible.

It is the right place to remind that dissipation (energy input) or drag only are not sufficient to define the properties of a turbulent flow. For example, Bevilaqua and Lykoudis (1978) performed experiments on flows past a sphere and a porous disc with the same drag. However, other properties of these flows even on the level of velocity fluctuations were quite different; see also Wygnanski et al. (1986) who performed similar experiments with a larger variety of bodies with the same drag, and also references in George (2012). It should be kept in mind that in both cases the flow was partly turbulent and most probably had different *large scale stability* properties for different bodies not directly related to the turbulent nature of the flow within the wake. This may contribute too to the differences in the observations. Similarly, many properties of turbulent flows with rough boundaries are not defined uniquely by their friction factor either (Krogstad and Antonia 1999).

The existence of such weak solutions of the Euler equations has not been proven. Even though a nontrivial and important question is about what happens in the zero-viscosity singular limit to quantities described by smooth physically meaningful solutions of the Navier–Stokes equations. This concerns not only the conserved quantities in Euler such as energy, but also key quantities such as strain, vorticity, their production, acceleration (especially its solenoidal part) among others. More problematic are geometrical issues even the most elementary as the alignments of vorticity with the eigenframe of strain. In problems with long time behavior, i.e. infinite time limit there is a problem that the result, generally, appears not the same depending on the order which limit is taken first. This contrasts the absence of such a problem for any finite however large Reynolds number.

A variety of empirical observations show clearly that turbulence is not a slightly viscous/dissipative phenomenon and that inertial and dissipative effects are of the same order at whatever Reynolds number and are not limited each by its own range of scales. This is manifested not only by the finite energy dissipation rate, but among other evidence, for example, by the so called Tennekes and Lumley balance between the enstrophy production (and similarly strain) and its viscous destruction/'dissipation', so that vorticity is not frozen in the flow field—again at any, however, large Reynolds numbers. This is not surprising as the zero-viscosity limit is singular. Nevertheless, there are numerous attempts to construct "theories" based solely on Euler equations with justifications such as that *a theorem which is valid for any finite, but very large, Reynolds number is expected to be compatible with results concerning infinite Reynolds number* (Bardos and Titi 2007), also (Migdal 1995). Such expectations, however, are not convincing since as mentioned the vanishing viscosity limit of solutions of Navier–Stokes equations is singular. Moreover, the existing experimental evidence does not provide any indication whatsoever that Euler equations are relevant to turbulence which is a strongly dissipative phenomenon of non-local nature at any however large Reynolds number.

Another theme is about the so called truncated Euler equations in which *the thermalized modes* (smaller scales) *act as a fictitious microworld on modes with smaller wave numbers in such a way that the usual dissipative Navier–Stokes dynamics is recovered at large scales* because *artificial microscopic systems can act just like the real ones as far as the emergence of hydrodynamics is concerned* and because *dissipation in real fluids is just the transfer of macroscopically organized (hydrodynamic) energy to molecular thermal energy* (Frisch et al. 2008). One of the problems here is that in real fluids the role of strain field is not limited just by dissipation. The consequence is that the nature of dissipation makes an essential qualitative difference including nonlocal effects so that different dissipative mechanisms result in different outcomes including the "inertial range" contrary to the common beliefs, see Chaps. 7 and 8.

3.3 Averaged Equations, Filtering, Decompositions and Similar Approaches/Issues

Based on detailed assessment of an attractor dimension it is shown that a low dimensional quantitative model is very likely fools' quest (Sirovich 1997).

The intricacy of turbulent flows as a high dimension phenomenon with its many degrees of freedom and nonlinear not weak interaction prompted the question whether a kind of low-dimensional description is possible (Kraichnan and Chen 1989), for more see Chap. 3 in Tsinober (2009). It was made meaningful by the fortunate empirically observed property of turbulent flows exhibiting reproducible statistical properties including averages, though had nothing more justifying the assumption of such a possibility. This resulted in Reynolds averaged Navier–Stokes equations (RANS) and large eddy simulations (LES) and similar modeling approaches such as methods of stochastic modeling. The key word is "modeling" as all of these approaches produce "equations of turbulent motion", using the language of Monin and Yaglom (1971, p. 20), which are always unclosed. The consequence is that one needs additional information to close these equations. This is the so called closure problem, in which one treats explicitly only a small fraction of the whole flow field, e.g. modes in some decomposition and represent the dynamic effects of the "rest" of the flow, typically small scales, by some additional information based on dimensional analysis, variety of scaling arguments, symmetries, invariant properties and various assumptions, many of which are of unknown validity and obscured physical and mathematical justification (if any) along with using non-trivial "surgeries" with removal large fractions of the flow field. The really acute question is whether all these "equations of turbulent motion" are adequate in describing the genuine/true physics of turbulent flows, i.e. whether all they do really deserve to be termed as "equations of turbulent motion" or they just mimic some the properties of turbulent flows not necessarily for the right reasons. It is only formally the origin of the closure problem looks as essentially technical due to the nonlinearity, but formidably difficult and so far unsolved, if ever. The above questions are of conceptual nature and utmost importance as they are about the nature, properties and the role of the "rest", i.e. of the part of the flow not treated explicitly, i.e. modeled.[3] In other words, these questions are about whether or not adequate reduced description of the highly dimensional problem is possible from the fundamental point of

[3]Saffman wrote in (1968): *A property of turbulent motion is that the boundary conditions do not suffice to determine the detailed flow field but only average or mean properties. For example, pipe flow or the flow behind a grid in a wind tunnel at large Reynolds number is such that it is impossible to determine from the equations of motion the detailed flow at any instant. The true aim of turbulence theory is to predict the mean properties and their dependence on the boundary conditions.*

The latter view is still very popular in the community. Such an aim may be interpreted as giving up important aspects of understanding the physics of basic processes of turbulent flows. Indeed, there exist a multitude of various "theories" predicting at least some of the mean properties of some flows, but none seem to claim penetration into the physics of turbulence.

It may be said that most of the theoretical work on the dynamics of turbulence has been devoted (and still is devoted) to ways of overcoming the difficulties associated with the closure problem (Monin and Yaglom 1971, p. 9).

These difficulties have not been overcome and it does not seem that this will happen in the near future if at all. There are several reasons for this. One of the hardest is the nonlocality property which is discussed in Chap. 7.

We mention that formally there exist two closed formulations which in reality are suspect to be just a formal restatement of the Navier–Stokes equations, at least, as concerns the results obtained to date. One is due to Keller and Friedmann (1925) infinite chain of equations for the moments and the equivalent to this chain Hopf equation in term's of functional integrals (Hopf 1952).

view. The crucial—explicit and/or implicit—assumption here is that the "rest" is somehow 'slaved' to the explicitly treated part of the flow (ETPF) without any reaction back and serving at best as a passive sink of energy so that the 'rest' can be "parameterized" in a simple way. This is a major misconception: the rich direct an bidirectional coupling between large and small scales comprises an essential part of the complex interaction between the multitude of the degrees of freedom in turbulent flows. Indeed, there is a variety of manifestations of direct and bidirectional impact/coupling of large and small scales which is essentially nonlocality: turbulent flows appear to be far more nonlocal than a theoretician would like to encounter.

Indeed, even just looking at the equations for the small/unresolved scales it is straightforward to realize that the small/unresolved scales depend on the large/resolved scales via nonlinear space and history-dependent functionals, i.e. essentially non-local both spatially and temporally. So it is unlikely—and there is accumulating evidence for this—that relations between them (such as "energy flux", but not the only) would be approximately local in contradiction to K41a hypotheses and surprisingly numerous efforts to support their validity.

One of the severe consequences of low dimensional approaches, i.e. closure, is that the essential dynamics and physics of the rotational and dissipative effects, which are mostly "small scale" ones, are treated in terms of transport coefficients like an eddy viscosity.

Another related misconception is that the large/resolved, etc. scales obey in some sense just the Euler equation, e.g. Landau and Lifshits (1987), Migdal (1995).

As it stands now all the enormous effort in this direction—being useful in a great variety of applications—lacks inherently the basis to be used in the issues of fundamental nature, though some people believe that *successful approximation methods would almost certainly illuminate the physics of turbulent flow* (Salmon 1998). The key words are 'successful approximations' and the remaining big question is how successful are "equations of turbulent motion" in the sense just mentioned above in the context of basic physics of turbulence. So far, any modeling remains *modeling only* and in principle cannot contribute to the basic aspects of turbulence problem and our understanding of the (nature, etc.) of turbulence. Modeling still resembles what von Karman termed the "science of variable constants". The history of science shows clearly that the importance and non-trivial consequences of the fundamental research of practical importance far broader than for (validation of) LES, subgrid modeling, etc. cannot and should not be underestimated.

Modeling involves typically a kind of decomposition: means versus fluctuations as in Reynolds decomposition, resolved-unresolved in LES and a variety of other decompositions both formal and heuristic. But the issue of using decompositions in turbulent flows is far broader. It is the introduction of some decomposition which along with the nonlinearity results in a "cascade" with its properties depending on the properties of the decomposition. This makes the popular concept of cascade ill posed which along with other important issues calls for reminding the acute and generic problems arising from employing decompositions.

A final note is about RDT approximation and similar quasi-linear approaches which pretend (in both meanings) to be able addressing issues of genuine turbulence both the problem and the phenomenon. However, the latter is a strongly non-linear phenomenon and its fundamental issues do not seem to be amenable to quasi-linear methods.

3.3.1 On the Utility of Various Decompositions

It is still true today as long before that there exist no *proper analytical machinery to be used* (von Neumann 1949), in turbulence research. In other words, there are no adequate tools to treat turbulence either analytically or via other theoretical approaches such as attempts to construct statistical and/or other theories. Consequently, among the most common approaches are those borrowed from methods used in *linear* problems both in theory and data analysis in turbulence which are mostly reductionist ones. A typical example is represented by various decompositions of the flow field. Being useful in many respects in the analysis of both flow states and processes decompositions have a rich potential (i) to be misused leading to misconceptions and ambiguities and introducing non-trivial artifacts and spurious effects due to properties of the decompositions not characteristic of genuine turbulence and not found in physical space and (ii) become an object of study by themselves keeping many people busy to a large extent with the properties of decompositions themselves rather than with genuine physics of turbulence. The above is among the reasons why decompositions became to some extent less productive/useful and even obscuring the physics of turbulence. A typical example of rather popular engagement is the study, sometimes pretty sophisticated, of interaction of the components of some decomposition not necessarily reflecting any physics, at least as concerns physical space, though frequently claiming confirmation of *the classical energy cascade picture* such as in the latest examples in Aluie (2012), Leung et al. (2012) and references therein, see Chap. 7.

There is a multitude of decompositions from formal to heuristic ones and there are several difficulties with all/any decompositions as an essentially linear procedure mainly due to the nonlinear and nonlocal nature of turbulence. These difficulties are not trivial and are 'generic'. The outstanding examples are the ill defined concepts of cascade as born via introducing some decomposition with dependence on the "physics" of the specific decomposition, inertial range, large scale modeling, all with the nonlinearity considered as the main guilty though nonlocality is not less malignant.

By nonlocality is meant (among other things) the direct an bidirectional coupling between large (resolved) and small (unresolved) scales, see Chap. 6 in Tsinober (2009), and below Chap. 7. Indeed, the most accepted division of turbulent flows on large and small scales is to a large extent artificial and in some sense even unphysical due to strong coupling between the two and due to an ambiguity of the very term 'scale' and the problematic nature of the decomposition approach to the phenomenon of turbulence and the necessity to handle turbulence as a whole.

More specific issues involve the phenomenon of cancellations with the total being much smaller than contributions from '+' and '−' modes, e.g. (Meneveau 1991a,

1991b; Chen et al. 2003), and misinterpretations of the meaning of this "smallness" of the total; employing wavepacket filters and/or narrow local wavepacket filters with some results very similar in the NSE turbulence and the corresponding random Gaussian field, e.g. that "pancake eddies" in turbulent flows possess strong alignments between velocity and vorticity, the so-called local Beltramization of purely kinematic nature (Yakhot and Orszag 1987a, 1987b; Shtilman 1987); the property of the nonlocal interactions with local transfer character of triadic interactions as not a property of turbulence physics, but rather a general feature of the Fourier representation (Waleffe 1992); and the general kinematic effect of "narrow banding" leading to Gaussian statistics in the narrow bands of modes whatever the departure from Gaussian statistics in physical space.

Moreover, *with a large number of degrees of freedom, a large class of plausible phase-space distributions have low-order moments that are indistinguishable from the low-order moments of the Gaussian distribution. This result is just a special case of the central limit theorem* (Orszag 1977). See also Lumley (1970, pp. 50 and 91), Leonov and Shiryaev (1960).

These are examples of warning against *overprocessing* of the data.

A closely related is the hunt (mostly futile so far) for "simple basic structures" and other 'simple' objects usually embedded in a "structureless random" background, such as the so called "coherent structures" that "govern the dynamics of the flow" and similar strong claims though there seems to exist no simple dynamically relevant structure(s) of the kind hunted for quite while (worms, sheets, vortex structures/filaments, vortons, 'eigensolutions', significant shear layers, etc.). Those which are observed are very far from being simple, but frequently misinterpreted as such. This is among the reasons and one of the acute problems that the structure(s) is (are) not represented by the modes of any known formal decomposition, and heuristic attempts at representation of turbulence as a collection of simple objects. For example, a Fourier-decomposition of a flow in a box with periodic boundary conditions. The emergence of structures in such a flow, such as the slender vortex filaments, in a random fashion (at random times with random orientation, and to a large extent random shapes) points to the limitation of Fourier decomposition or similar, which does not 'see' these or any other real physical structure(s) which is generally true of any decompositions. Another example is the chaotic regime of a system with few degrees of freedom only, e.g. three as in the Lorenz system (Lorenz 1963), or four in the forced spherical pendulum (Miles 1984), but with a *continuous* spectrum. This points to a serious problem since components of a decomposition are not synonymous and cannot be automatically identified as degrees of freedom. Hence the problems with the ambiguity of decompositions. A noteworthy problem is that there is an inherent problem of definition of scales generally and in terms of some decomposition. A serious problem is that the objects claimed to represent "simple" elements of turbulence structure are "educed" from time snapshots of the flow field do not possess any (e.g. Lagrangian) identity allowing to follow their time evolution unambiguously in a reasonable manner except of "cartoon-like" approaches and alike. The problem is more general, e.g. vortex lines having no Lagrangian identity as it is not frozen in the flow field at however large Reynolds numbers and consequently problems with the phenomenon of reconnection.

A recent (but not the only) example concerns employing band pass filters decomposition. On one hand Eyink and Aluie (2009a, 2009b) insist on *localness of energy cascade*, i.e. *that the interscale energy transfer is dominated by local triadic interactions.* On the other hand Leung et al. (2012) find quite a bit of non-locality in the relation between filtered vorticity and strain (*Most of the enstrophy generation occurs during stretching by the largest strain... due to non-local straining*) but at the same time claiming that they *have found that the classical view of the energy cascade is qualitatively correct, with energy passing down the cascade as large vortices straining the smaller ones.* Perhaps, it is the right place to remind that the problematic concept "cascade" whatever this means, is a local process by definition.

A most popular paradigmatic heuristic example is the decomposition on energy-containing (ECR), inertial (IR) and dissipative ranges (DR). It is massively accepted/believed that the statistical properties of IR at large Reynolds numbers are universal (in some sense) and independent of viscosity/nature of dissipation and consequently of the properties of DR. In particular, it is widely believed that *Kolmogorov's basic assumption* (Kolmogorov 1941a) *is essentially that the internal dynamics of the sufficiently fine-scale structure (in x-space) at high Reynolds numbers should be independent of the large-scale motion. The latter should, in effect, merely convect, bodily, regions small compared to the macro scale* (Kraichnan 1959). This is what is called sweeping which is claimed to have purely kinematic nature. Though these hypotheses indeed appear to be approximately correct due to the kinematical effects, the subtle point is that they are erroneous conceptually as missing essential effects responsible for all the dynamics in the Eulerian representation, see Sect. 7.3.

3.4 Eulerian Versus Lagrangian Representations

Though these are formally equivalent the majority of the issues are considered in Eulerian setting in which the observation of the system is made in a fixed frame as the fluid goes by. In this case the motion is characterized by the velocity field $\mathbf{u}(\mathbf{x}, t)$ as a function of position vector, $\mathbf{x}$, and time, t. In the Lagrangian[4] setting the observation is made following the fluid particles wherever they move. Here the dependent variable is the position of a fluid particle, $\mathbf{X}(\mathbf{a}, t)$, as a function of the particle label, a (usually it's initial position, i.e., $\mathbf{a} \equiv \mathbf{X}(0)$) and time, t. With the relation between the two ways of description[5] both are formally equivalent. However, there are essential differences of conceptual nature rather than just technical ones. As concerns the technical side, the NSE equations in a pure Lagrangian (LNSE) setting are intractable so far for viscous flows. However, the very form of LNSE as compared to ENSE provokes nontrivial and important questions of conceptual nature.

[4]In fact it is also due to Euler, see Lamb (1932). A detailed account on the 'misnomer' by which the 'Lagrangian' equations are ascribed to Lagrange is found in Truesdell (1954, pp. 30–31).

[5]This relation is given by the equation $\partial\mathbf{X}(\mathbf{a}, t)/\partial t = \mathbf{u}[\mathbf{X}(\mathbf{a}, t); t]$, i.e., the Lagrangian velocity field, $\mathbf{v}(\mathbf{a}, t) = \partial\mathbf{X}(\mathbf{a}, t)/\partial t$, is related to the Eulerian velocity field, $\mathbf{u}(\mathbf{x}, t)$, as $\mathbf{v}(\mathbf{a}, t) \equiv \mathbf{u}[\mathbf{X}(\mathbf{a}, t); t]$.

The first issue of this kind is that in the Lagrangian setting the fluid particle acceleration is linear in the fluid particle displacement so that the 'inertial' nonlinear effects are manifested only by the term containing pressure, i.e. there are no terms like the advective terms $(\mathbf{u} \cdot \nabla)\mathbf{u}$ in the pure Eulerian setting and consequently no sweeping. Moreover, one can hardly speak about things like Reynolds decomposition and Reynolds stresses, turbulent kinetic energy production in shear flows in a pure Lagrangian setting. The inertial, i.e. non-viscous nonlinearity in the Lagrangian representation cannot be interpreted in terms of some cascade as it cannot be maintained solely by pressure gradient, and it is far less clear how one can employ decompositions even at the problematic level as done in the pure Eulerian setting.

Another issue of conceptual nature is comes from the seemingly simple kinematic equation relating the Eulerian and Lagrangian settings. This equation is nonlinear for almost all even for very simple fluid flows and is generically nonintegrable for all flows with the exception of very simple flows such as unidirectional ones. Thus for a wide class of almost all laminar flows in the Eulerian setting, i.e. with the Eulerian velocity field, $\mathbf{u}(\mathbf{x}; t)$ not chaotic, regular and laminar, the Lagrangian velocity field $\mathbf{v}(\mathbf{a}, t) \equiv \mathbf{u}[\mathbf{X}(\mathbf{a}, t); t]$ as any other property of fluid particle is chaotic because $\mathbf{X}(\mathbf{a}, t)$ is chaotic! This fact is of utmost importance for issues like the relation between the Eulerian and Lagrangian (statistical) characteristics of the same flow field. It has to be emphasized that this chaotic behavior is of purely kinematic nature resulting solely from the equation relating the Eulerian and Lagrangian settings and has nothing to do with dynamics, i.e. genuine (as NSE) turbulence. This concerns also all problems with prescribed velocity fields in Eulerian setting—synthetic, Gaussian, etc. Similarly, all randomly forced E-laminar flows with low Reynolds number including multiscale ones, flows in porous media, microdevices, to name some, belong to this category.

Whereas the E-turbulence is a dynamic phenomenon this is not necessarily the case with the L-turbulence which may be a purely kinematic one. In other words the Eulerian setting reveals the pure dynamic chaotic aspects of genuine turbulence as contrasted to "mixing" of the kinematical with the dynamic ones in the Lagrangian setting, i.e., in genuine turbulence the latter contains both which seem to be essentially inseparable. Consequently, studying Lagrangian statistics only may not provide adequate information of the L-statistics of genuine turbulence as not necessarily containing its pure dynamic "stochasticity". In other words the flow can be purely L-turbulent, i.e. E-laminar as mentioned above. However, if the flow is E-turbulent, i.e. $Re \gg 1$ it is L-turbulent as well. One more consequence is that the structure and evolution of passive objects including fluid particles in genuine turbulent flows arises from two contributions: one due to the Lagrangian chaos and the other due to the complex nature of the Eulerian velocity field itself.

Thus the ENSE and LNSE are not equivalent as the latter appear to be 'more chaotic' with the multitude of E-laminar flows having no counterpart to the corresponding L-turbulent statistics in the same fluid flow. The equivalence is between the LNSE with the ENSE plus the equation relating the Eulerian and Lagrangian settings.

3.5 Final Remarks

Returning to the NSE, the conclusion that NSE are useful as only an experimental tool would be incorrect. It is true that there is little substantial theoretical use of NSE in turbulence and the road to high *Re* does not look as short even in the simplest geometries. However, there are several ways to do this implicitly, i.e., by indirect use of NSE and their consequences. For example, looking at the NSE and their consequences themselves enables one to recognize at least some of the dynamically important quantities and physical processes involved including some important nonlinear objects many of which still await to be identified and properly defined. In other words, NSE and their consequences tell us what quantities and relations should be studied aiding the difficult task of identifying the significant nonlinear objects of turbulence. So far this can be done mostly experimentally,[6] but this kind of 'guiding' is hoped to be also useful theoretically. The most elegant exception is a set of theoretical results on the a priory upper bounds of long time averages of dissipation and global transport of mass, momentum and heat, see Doering (2009) and references therein.

It looks that in order to solve the standard NSE problems, there is a need for progress in the understanding of turbulence as a physical phenomenon and not the other way around. So there is every reason to follow the advise of Kolmogorov (1985) to go on with non-trivial observations. The role of mathematics in physical sciences cannot be underestimated. As concerns turbulence mathematics it is supposed to aid the goals of basic physical problems. However, in case of turbulence mathematics did not became what von Neuman and von Karman would expect and played a pretty modest role especially as concerns the qualitative physics of turbulence mainly limited by proofs of well known facts and related via employing methods of functional analysis with Sobolev, Besov and other more exotic spaces with not less exotic objects. These are "suspect" to belong to excessive generalizations or 'overgeneralizations' dealing with objects of unknown relevance to the physical phenomenon of turbulence and unjustified claims that sometimes it makes physical sense to admit solutions with singularities in spite that, generally, singularities are not admissible from the physical point of view and, in fact, nobody can tell so far what is the relation between the two. The big problem is that it is not at all clear how the results obtained with these singular (!) objects, e.g. distributions are relevant/related or even have anything to do with turbulence. The issue is broader and extends well into the theoretical physics either.

This tendency for overgeneralization in science was criticized by Poincare at the beginning of the 20-th century 1902:

And in demonstration itself logic is not all. The true mathematical reasoning is a real induction, differing in many respects from physical induction, but, like it, proceeding from the particular to the universal. All the efforts that have been made to upset this order, and to reduce mathematical induction to the rules of logic, have

[6] Victor Yudovich used to say that one of the best solutions of NSE is experiment (1971, private communication).

ended in failure, but poorly disguised by the use of a language inaccessible to the uninitiated (Poincare 1952a).

Some have set no limits to their generalizations, and at the same time they have forgotten that there is a difference between liberty and the purely arbitrary. So that they are compelled to end in what is called nominalism; they have asked if the savant is not the dupe of his own definitions, and if the world he thinks he has discovered is not simply the creation of his own caprice (Poincare 1952b, pp. xxiii–iv).

For example, Foiaş et al. (2001) have shown that there are measures on a function space that are time-invariant. However, invariance under time evolution is not enough to specify a unique measure which possibly would describe turbulence. An attempt to select the right one was made by Ruelle (1979): *The effect of thermal fluctuations on a turbulent flow is estimated, and it is argued that these fluctuations are important in selecting the stationary measure on phase space which describes turbulence.*

Another problem is that it is not clear how the objects that the authors have constructed and used in their proofs are relevant/related or even have anything to do with turbulence.

All these and similar ones are of interest for their own sakes, but from the point of view of theoretical study of turbulence and its understanding it is definitely not the first priority to reproduce in a (more or less) rigorous manner simple known results by complicated theory. *It may be original to produce a simple known result by a complicated theory, but mathematical study of turbulence is not supposed to have its main aim as providing full employment for abstract mathematicians* (Saffman 1978). Quoting von Karman (1943), *in order to get the solution of engineering problems..., you need some kind of tool designers. These are the real applied mathematicians. Their original backgrounds may differ..., but their common aim is to "tool up" mathematics for engineering* and physics. It is of interest to contrast the statement by Friedrichs: *Applied mathematics is those areas in which physicists are no longer interested.* In turbulence there seem to be still no such tool up at all. Mathematicians mostly live their own life in a variety of functional spaces and even have their own Journal of Mathematical Fluid Dynamics. However, this is only a part of the problem. *Sometimes experiments provide us with so beautiful and clear results that it is a shame on theorists that they cannot interpret them* (Yudovich 2003), not to say to guide the observations. I am afraid that not many care about the observational evidence. The approach in reality is in some sense reverse. The widespread view of both mathematicians and theoretical physicists is that the main function of all experiments/observations both physical and numerical is to "validate theories"—paradoxically nonexistent so far. This is beautifully reflected by Monin and Yaglom (1971, p. 21) especially in the last sentence of the following quotation: *The combination of theoretical and experimental approaches, which is extremely fruitful in all investigations of natural phenomena, is especially necessary in statistical fluid mechanics where the theory is often still of a preliminary nature and is almost always based on a number of hypotheses which require experimental verification.* ***However, we have avoided introducing experimental results which have no theoretical explanation and which do not serve as a basis for some definite***

theoretical deductions, even if these data are in themselves very interesting or practically important. It is an approach which to a large degree what not unjustly was called "postdiction" and is not useful especially in a situation in which from the basic point of view there is almost nothing to be confirmed. *Correlations after experiments done is bloody bad. Only prediction is science* (Hoyle 1957). In other words, the role of experiments in turbulence goes far beyond the view of those who think of experimentalists as a superior kind of professional fixers knowing how to turn nuts and bolts into a confirmation of other people's 'theories'. The issue is more than serious in view of the general theoretical failure, i.e. absence of theory based on first principles. There is a number of issues of special concern about the relation of what is called theory in turbulence research and observations/experiments, in particular, regarding the use of the factual information as concerns fundamental aspects.

Part II
Issues of Paradigmatic Nature I: Origins and Nature of Turbulence

The true problem of turbulence dynamics is the problem of its origin(s) and successive development from some initial and with some boundary conditions and forcing to an ultimate state assuming "for simplicity" that this state is statistically stationary, i.e. with statistically stationary forcing. However, since this route is extremely complicated and involved, a second approach is used quite frequently using statistical methods with the price being of loosing important aspects of dynamics among other. Namely, turbulent flows are studied 'as they are' disregarding their origin. This was prompted also by the beliefs in universality and lack of "memory" as concerns the initial and boundary conditions and the nature of the mechanisms of turbulence production and sustaining. It appears that there is far less universality and quite a bit of memory than expected. In both approaches, at least some aspects of the time evolution are of central importance, since turbulence dynamics is a process and has memory along with other properties such as nonlocality with its diverse manifestations. These are the reasons that before getting to the main theme—the fully developed turbulent flows—an overview is given of the origins of turbulence. This is followed by the discussion of the various facets of the "undeniably statistical nature", e.g. (Foiaş et al., Navier–Stokes equations and turbulence, 2001) contrasted to the "deterministic origin" of turbulence as described by purely deterministic non-integrable equations and related issues, such as the necessity of replacing the paradigm of the meaning of apparent randomness, stochasticity of turbulence which is roughly just the complexity due to a large number of strongly interacting degrees of freedom governed by purely deterministic equations. These comprise the first part of the major qualitative universal features of turbulent flows briefly described/listed in the Introduction. Considerable attention is given to the limitations of statistical methods.

Chapter 4
Origins of Turbulence

Abstract It is a common view (but not the only) that the origin of turbulence is in the instabilities of some basic laminar flow. As the Reynolds number increases, some instability sets in, which is followed by further instabilities/bifurcations, transition and a fully developed turbulent state and the processes by which flows become turbulent are quite diverse. One process deserves special attention as a specific universal phenomenon. It is the continuous transition of laminar flow into turbulent via the entrainment process through the boundary between laminar/turbulent regions in the partly turbulent flows. In all these there is a distinct Lagrangian aspect: the abrupt transition of fluid particles (i.e. Lagrangian objects) from the laminar to turbulent state when passing across the laminar/turbulent 'interface'. Abrupt is a key word, i.e. without any cascade whatsoever. Another issue of special interest is that there is a conceptual difference between the two kinds of flows: those arising 'naturally', e.g. by a simple time independent and smooth in space deterministic forcing, i.e. governed by a purely deterministic system and those produced by some external random source. The point is illustrated by the examples of stochastic forcing of integrable nonlinear equations (Burgers, Korteveg de Vries, etc.) which without such forcing do not exhibit any "stochastic" behavior whatsoever. In other words, the properties of a turbulent flow are not neutral to the nature of excitation. Likewise boundary, initial and inflow conditions may cause qualitative differences either especially due to essential role of nonlocal properties of the system.

There is a qualitative difference between transition to turbulence as a phenomenon characterized by a large number of strongly interacting degrees of freedom and transition to chaotic behavior, in general, and between such notions as degrees of freedom and the dimension of attractor (assumed to exist), in particular. Any fluid flow which is adequately represented by a low-dimensional system is not turbulent—a kind of definition of 'non-turbulence'. The immediate examples are laminar in Euler setting (E-laminar) low-dimensional chaotic fluid flows (L-turbulent).

The state of some particular flow configuration depends on the magnitude of the Reynolds number, *Re*, but is not solely defined by *Re*. At small enough *Re* the flow is laminar for any initial conditions and is the same for stationary conditions. At larger *Re* the flow state, generally, depends strongly of initial conditions, at large

A. Tsinober, *The Essence of Turbulence as a Physical Phenomenon*,
DOI 10.1007/978-94-007-7180-2_4,

Re is turbulent and is believed to be smooth enough and unique in some sense at any *Re*. For example, for statistically stationary conditions the asymptotic state is an attractor—a complex non-trivial object (Doering and Gibbon 2004; Foiaş et al. 2001; Robinson 2001, 2007).

4.1 Instability

It is a common view (but not the only) that the origin of turbulence is in the instabilities of some basic laminar flow. This is understood in the sense that any flow is started at some moment in time from rest, and as long as the Reynolds number or a similar parameter is small, the flow remains laminar. As the Reynolds number increases, some instability sets in, which is followed by further instabilities/bifurcations, transition and a fully developed turbulent state. Such sequences of events occur, generally, not throughout the whole flow field, but at successive downstream locations in spatially developing flows, at the laminar/turbulent 'interfaces' in turbulent spots and in all partly turbulent flows which comprise the majority of all observed ones. The main special features of these flows are the coexistence of regions with laminar and turbulent states of flow and continuous transition of laminar flow into turbulent via the entrainment process through the boundary between the two. In all these there is a distinct Lagrangian aspect: the abrupt transition of fluid particles (i.e. Lagrangian objects) from the laminar to turbulent state when passing across the laminar/turbulent 'interface'. Abrupt is a key word, i.e. without any cascade whatsoever.

From the mathematical point of view the transitions from one flow regime to another with increasing Reynolds number—as we observe them in physical space—are believed to be a manifestation of generic structural changes of the mathematical objects called phase flow and attractors in the phase space through bifurcations in a given flow geometry (Hopf 1948), though partly turbulent flows with the special feature of these flows being the coexistence of regions with laminar and turbulent states of flow are not easily 'fit' in this picture.

Whereas the processes by which flows become turbulent are quite diverse,[1] most known qualitative and some quantitative properties of many (but not all) turbulent

[1]The diversity of the processes by which flows become turbulent is in part due to the sensitivity of the instability and transition phenomena to various details characterizing the basic flow and its environment. For example, the Orr-Sommerfeld equation governing the linear(ized) (in)stability contains the second derivative of the basic velocity profile. Many flows (some of the so-called open flows) are very sensitive to external noise and excitation. There are essential differences in the instability features of turbulent shear flows of different kinds (wall bounded—pipes/channels, boundary layers, and free—jets, wakes and mixing layers), thermal, multidiffusive and compositional convection, vortex breakdown, breaking of surface and internal waves and many others. It is important that such differences occur also for the *same* flow geometry, which display in words of M.V. Morkovin *bewildering variety of transitional behavior*. The specific route may depend on initial conditions, level of external disturbances (receptivity), forcing, time history and other details in most of the flows mentioned above. This diversity is especially distinct for the very initial stage—the (quasi)linear(ized) instability. Later nonlinear stages are less sensitive to such details.

flows were considered as do not depending either on the initial conditions or on the history and particular way of their creation, e.g. whether the flows were started from rest or from some other flow and/or how fast the Reynolds number was changed. However, relatively recent evidence shows that there exist differences deserving special attention; some flow properties, generally, do depend on the nature of excitation, i.e. forcing and inflow, initial and boundary conditions. This calls for a discussion and reexamination of such issues as universality, symmetries and nonlocal properties of turbulent flows among others.

One of the basic features of utmost importance of processes resulting in turbulence is that all of them tend to enhance the field of velocity derivatives, i.e. the rotational and dissipative properties of the flow in the process of transition to turbulence. The first property is associated with the production of vorticity (i.e. its rotational nature), whereas the second property is due to the production of strain, i.e. its dissipative nature. These two key properties of all turbulent flows are among our main concerns below.

4.2 Transition to Turbulence Versus Routes to Chaos

One of the main achievements of modern developments in deterministic chaos is the recognition that chaotic behavior is an intrinsic fundamental property of a wide class of nonlinear physical systems (including turbulence) described by purely deterministic equations/laws, and not a result of external random forcing or errors in the input of the numerical simulation on the computer or the physical realization in the laboratory. The nonlinear systems and the equations describing them produce an apparently random output 'on their own', 'out of nothing'—it is their very nature. However, there is a variety of qualitatively different systems exhibiting such a behavior just like there is a large diversity of such behaviors.

The qualification of turbulence as a phenomenon characterized by a large number of strongly interacting degrees of freedom enables us to make a clear distinction between transition to turbulence and transition to chaotic behavior, in general, and between such notions as degrees of freedom and the dimension of attractor (assumed to exist), in particular.

The main points of distinction are as follows.

Low dimensional chaotic systems like the famous Lorenz (1963) system or the spherical pendulum studied by Miles (1984) change their behavior from simple regular (as periodic) to distinctly chaotic as some parameter of the system changes. However, obviously the number of degrees of freedom of all such systems remains the same, only the character of the interaction of these degrees of freedom changes. In contrast, the idea that the essential feature of transition to turbulence[2] is an increase of the number of excited degrees of freedom dates back to Landau (1944)

Hence there is some tendency to universality in strongly nonlinear regimes, such as developed turbulence.

[2]Not to be confused with transition to low dimensional (usually) temporal chaos.

and Hopf (1948) and is correct, though the details of their scenario appeared to be not precise, see Monin (1986). However, *Kolmogorov's ideas on the experimentalist's difficulties in distinguishing between quasi-periodic systems with many basic frequencies and genuinely chaotic systems have not yet been formalized* (Arnold 1991). In other words it is very difficult if not impossible to make such a distinction in practice.

It is now recognized that *despite the considerable successes of the present studies of the application of modern ideas on chaos to well-controlled fluid flows, they appear to have little relevance when applied to the more general problem of fluid turbulence* (Mullin 1993, p. 93); see also Tritton (1988, p. 410) and that *considerations of the properties of fully-developed turbulence require rather different ideas* (Batchelor 1989).

So it is quite plausible that any fluid flow which is adequately represented by a low-dimensional system is not turbulent—a kind of definition of 'non-turbulence'. The immediate examples are laminar in Euler setting (E-laminar) low-dimensional chaotic fluid flows (L-laminar) and other examples mentioned in the introduction above and Chaps. 3 and 9 in Tsinober (2009).

Here is the right place to note that there is an important difference between the number of degrees of freedom roughly proportional to the number of ordinary differential equations necessary to adequately represent a system described by partial differential equations (NSE) and the dimension of the attractor of the system, though sometimes it is considered *natural to identify the dimension of the attractor as an effective measure of the number of degrees of freedom in the system* (Doering and Gibbon 2004). However, in a particular dynamic system described by ODE, the former is obviously fixed and is independent of the parameters of the system, whereas the latter is changing with the parameters but is bounded. In turbulence both are essentially increasing with the Reynolds number and become very large at large Reynolds number.

4.3 Many Ways of Creating (Arising/Emerging) Turbulent Flows

Any turbulent flow is maintained by an external source of energy produced by one or more mechanisms. The mechanisms maintaining/sustaining and influencing turbulence, at least some of them, are believed to be closely related (but are not the same) to those by which laminar and transitional flows become turbulent. Apart from a great variety of turbulent flows in nature/technology and 'natural' ways resulting from instabilities, turbulent flows can be produced by 'brute force', i.e. by applying external forcing of various kinds both in real physical systems and in computations by adding some forcing in the right hand side of the Navier–Stokes equations. For example, one of the "simplest" kinds of turbulent flow—quasi-homogeneous and quasi-isotropic—can be established by moving a grid through a quiescent fluid or placing such a grid in a wind tunnel, or oscillating such grids in a water tank. Turbulence is produced by forcing in the interior of the fluid flow (by electromagnetic forces, e.g. in electrolytes, liquid metals or plasma; or other body forces) or at flow boundaries, which can be still or moving/flexible, smooth or rough, simple or complex. Similarly turbulent flow can be produced numerically with an infinite versatility by adding a force (random or deterministic) to the right hand side of the Navier–Stokes equations and/or forcing the flow at its boundaries.

An important point is that the nature of forcing (deterministic, random, temporally modulated or whatever) is secondary in establishing and sustaining a turbulent flow, provided that the Reynolds number is large enough and the forcing is mostly in the large scales, but, in contrast to the common beliefs, not necessarily unimportant to the particular properties of the resulting turbulent flow. Moreover, there is a conceptual difference between the two kinds of flows: those arising 'naturally', e.g. by a simple time independent and smooth in space deterministic forcing, i.e. governed by a purely deterministic system and those produced by some external random source.[3] The point is illustrated by the examples of stochastic forcing of integrable nonlinear equations (Burgers, Korteveg deVries, etc.) which without such forcing do not exhibit any "stochastic" behavior whatsoever. In other words, one can wonder about the differences between the turbulent flows produced by some stochastic forcing of NSE (or a real physical system) and those arising intrinsically, though it is a common practice to consider both as similar, at least qualitatively, so that the stochastically forced ones are even seen as representative of real turbulent flows. However, the properties of a turbulent flow are not neutral to the nature of excitation: if, e.g. it is not large scale may result in qualitative differences such as in case of broadband forcing.[4] Similarly boundary, initial and inflow conditions may cause qualitative differences either especially due to essential role of nonlocal properties of the system.

At small enough Reynolds numbers, the flow produced by deterministic forcing is not random, it is laminar, but the flow produced by random forcing, though random, is in many respects trivial as any randomly forced linear system, e.g. there is no interaction between its degrees of freedom. Strictly speaking this is true of pure dynamic flow properties described in Eulerian setting because some of its 'kinematic' properties as described in pure Lagrangian setting can be and usually are pretty complex and not trivial (L-turbulent), see, e.g. Fig. 1.3.

Thus a turbulent flow originates not necessarily out of a laminar flow with the same geometry. It can arise from any initial state including a 'turbulent' one, such as random initial conditions in direct numerical simulations of the Navier–Stokes equations. That is, the transition from laminar to turbulent regime is not the only causal relation. This problem is related to a somewhat 'philosophical' question on whether flows become or whether they just are turbulent, and to the unknown properties of the phase flow, attractors and related matters, which are mostly beyond the scope of this book.

---

[3]The random force method in turbulence theory is due to Novikov (1963).

[4]In this case if the forcing is strong enough not only in the large scales it can balance the viscous effects directly thereby bypassing the nonlinearity.

Chapter 5
Nature of Turbulence

Abstract The main dispute about the origins and nature of turbulence involves a number of aspects and issues in the frame of the dichotomy of deterministic versus random. In science this dispute covers an enormous spectrum of themes such as philosophy of science, mathematics, physics and other natural sciences. Fortunately, we do not have to dwell into this ocean of debate and opposing and intermediate opinions. This is mainly because (as it stands now) turbulence is described by the NSE which are purely deterministic equations with extremely complex behavior enforcing use of statistical methods, but this does not mean that the nature of such systems is statistical in any/some sense as frequently claimed. The bottom line is that turbulence is only apparently random: the apparently random behavior of turbulence is a manifestation of properties of a purely deterministic law of nature in our case adequately described by NSE. An important point is that this complex behavior does make this law neither probabilistic nor indeterminate.

One of the problems of turbulent research is that we are forced to use statistical methods in one sense/way or another. All statistical methods have inherent limitations the most acute reflected in the inability of all theoretical attempts (both physical and mathematical) to create a rigorous theory along with other inherent limitations of handling data such as description and interpretation of observations. However, the technical necessity of using statistical methods is commonly stated as the only possibility in the theory of turbulence. The consequence of this leads to the necessity of low-dimensional description with the removal of small scale and high-frequency components of the dynamics of a flow including quantities containing a great deal of fundamental physics of the whole flow field such as rotational and dissipative nature of turbulence among others. Thus, relying on statistical methods only (again with all the respect) one is inevitably loosing/missing essential aspects of basic physics of turbulence. So one stays with the troublesome question whether it is possible to penetrate into the fundamental physics of turbulence via statistics only. In other words, there is an essential difference between the enforced necessity to employ statistical methods in view absence of other methods so far and the impossibility in principle to study turbulence via other approaches. This is especially discouraging all attempts to get into more than just "en masse". Also such a standpoint means that there is not much to be expected as concerns the essence of turbulence using exclusively statistical methods.

A. Tsinober, *The Essence of Turbulence as a Physical Phenomenon*,
DOI 10.1007/978-94-007-7180-2_5,

Typical questions are as follows. Is turbulence just probabilistic or just deterministic? or both or whatever? Can it be just statistical? But perhaps, as it stands today the first question is whether these questions are as meaningful as thought before.

The main dispute about the origins and nature of turbulence involves a number of aspects and issues in the frame of the dichotomy of deterministic versus random. The latter term is frequently replaced by its synonyms such as statistical, stochastic, probabilistic: to stress this point we will use deliberately all of these terms as having the same meaning. In science this dispute covers an enormous spectrum of themes such as philosophy of science, mathematics, physics and other natural sciences.

Fortunately, we do not have to dwell into this ocean of debate and opposing and intermediate opinions.[1] This is mainly because (as it stands now) turbulence is described by the NSE which are purely deterministic equations.[2] It is due to this fact along with the developments in what is called deterministic chaos that the quite common contraposing the 'traditional' statistical and deterministic (and the so called structural) approaches in turbulence research has lost most (but not all) of its meaning. As mentioned it is well established that even simple systems governed by purely deterministic nonlinear sets of equations, such as those described by only three nonlinear ordinary differential equations, as a rule exhibit irregular apparently random/stochastic behavior. In fact, in respect to turbulence this was known long before the 'discovery of chaos': this deterministic system was (and is) studied by exclusively statistical methods, which are used today to study the statistical properties of various chaotic dynamical systems described by purely deterministic equations, e.g. Ornstein and Weiss (1991), Loskutov (2010).

Even the simplest nonlinear systems exhibiting chaotic behavior are analyzed via various statistical means. Also, the so called 'coherent structures' in turbulent flows are looked for using essentially statistical methods, such as conditional statistics though with limited success. Finally, methods of dimensional analysis, similarity and symmetries including group theoretical methods and phenomenological arguments are applied exclusively to quantities expressing the statistical properties of turbulent flows, see, e.g. Bonnet (1996), Holmes et al. (1996).

One can study the properties of deterministic systems by statistical methods, but this does not mean that the nature of such systems is statistical in any/some sense as frequently claimed.

---

[1] *Quantum mechanics is certainly imposing. But an inner voice tells me that it is not yet the real thing. The theory says a lot, but does not really bring us any closer to the secret of the "old one." I, at any rate, am convinced that He does not throw dice* (Einstein 1926).

Not only does God definitely play dice, but He sometimes confuses us by throwing them where they can't be seen (Hawking and Penrose 1996).

Today "chaotic" and "determinstic" are not considered as counterparts of a false dichotomy. This takes it origin to Poincare (1952b, pp. xxiii–iv).

[2] However, we repeat that since Leray (1934) until recently one was not sure about the theoretical, but not observational, possibility that turbulence is a manifestation of breakdown of the Navier–Stokes equations. Also note the statement by Ladyzhenskaya (1969): *...it is hardly possible to explain the transition from laminar to turbulent flows within the framework of the classical Navier–Stokes theory.*

5.1 Turbulence is Only Apparently Random

One of the most important aspects is that the apparently stochastic/random behavior of turbulent flows is their intrinsic property—as in intrinsic chaos—though in many other respects the NSE turbulence is qualitatively different. This is in contrast to stochastic NSE with the external random forcing or other excitation. There is no necessity for external random forcing either in the interior of the fluid flow or at its boundaries, nor does one need to start the turbulent flow with some random initial conditions provided that the Reynolds number is large enough. The fascinating popular question is how such a behavior emerges from purely deterministic equations as the Navier–Stokes equations are, deterministic forcing along with smooth nonrandom initial and boundary conditions. The common answer is the extreme sensitivity to disturbances whatever small (initial conditions, boundary conditions, external noise). However, this has little to do with complete randomness associated with the "absence of laws" as turbulence is governed by purely deterministic equations. The essential point and the bottom line is that *to say that turbulent flow is 'completely random' would define turbulence out of existence* (Tritton 1988).

The term "apparently" is stressed again for several reasons. The first—already mentioned—is that turbulence is described by purely deterministic equations. Second, turbulence only looks and exhibits seemingly/effectively random behavior. This extremely intricate and complex behavior stems from the non-trivial strong nonlinear and nonlocal interaction of an enormously large number of degrees of freedom (DOF) with the number of DOF increasing from unity for a stationary laminar flow at subcritical Reynolds number and growing as $\sim Re^{9/4}$ at large Reynolds numbers (Landau and Lifshits 1959), not later editions, since this estimate was removed from the subsequent editions. Another and a different way to put this is that what is called intrinsic stochasticity is just the *complexity of the dynamics along the attractor rather than its stability* (Arnold 1991), with highly non-trivial structure of the attractor governed by the NSE. The third concerns the very meaning of the term "random". Mathematicians and physicists alike have found it advantageous to introduce axiomatically the concept of complete randomness associated with the "absence of laws", which is not the case here.

Indeed, *Most problems in classical stochastic processes are reduced to solubility by statistical independence, or the assumption of a normal distribution (which is equivalent) or some other stochastic model; because of the governing differential equations, the turbulent velocity at two space-time points is, in principle, never independent—in fact, the entire dynamical behavior is involved in the departure from statistical independence. The equations, in fact, preclude the assumption of any ad hoc model, although this is often done in the absence of a better idea* (Lumley 1970).

It is the intricacy and complexity which is taken frequently as identical/synonymous to random/statistical nature of turbulence and which enforces to employ the statistical methods which is a quite different matter. The necessity and justification of using statistical methods comes both from our ignorance/lack of information about some aspects of the phenomenon in question—or, in words of

Kolmogorov (1956)—*a large number of random factors* and/or from excessively large amounts of information and consequently inability or lack of desire to handle this huge information otherwise ideally in full. The two situations are essentially the same as in both only partial information is used. In other words the probabilistic approach *is related, in part to this ignorance, in part to our knowledge* (Laplace 1951). The fascinating empirical outcome in both cases is exhibited by statistically stable properties, but the problematic aspect is that statistical methods both in theory and description and interpretation of the data from laboratory, field and numerical experiments are inherently limited/deficient, see next subsection.

On the other hand, the apparently random behavior of turbulence is a manifestation of properties of a purely deterministic law of nature in our case adequately described by NSE. An important point is that this complex behavior does make this law neither probabilistic nor indeterminate. Frequently assumed unpredictability of individual realizations of turbulent motion—which has made the application of stochastic theory attractive—is incompatible with the fact that this as any other realization is governed by NSE. The lack of predictability is related to what we human beings are able to observe, analyze and compute, i.e. in a sense subjective, independently whether this lack of predictability is due to sensitive dependence on initial conditions or not. The real issue here is that the law of nature we deal with is deterministic rather than probabilistic. After all it is well known that 'chaotic' (not synonymous with random!) behavior is fully compatible with deterministic laws. Thus the above issue is not identical to what is called "uncertainty", e.g. *Within classical physics, with their deterministic and precisely known laws, the evolution of many systems is nevertheless uncertain because these laws are chaotic* (Palmer and Hardaker 2011).

In discussing the nature of turbulence an important distinction has to be made between the nature of the phenomenon and of the problem. Indeed, one can assign statistical nature to a theory which is then qualified as statistical hydrodynamics claimed by many as synonymous to theory of turbulence or even mathematical turbulence theory, which is not, as—we repeat again—some aspects of turbulence can be treated via statistical methods, but it is a physical phenomenon which cannot be qualified as having purely statistical nature.

In fact, the "statistical nature" is introduced into the problem of turbulence at the outset "by hand" as the NSE are not stochastic PDE, they are purely deterministic. This is done in different ways and motivations. For example, just by claiming the necessity due to observations and treating the variables as random, e.g. *It is natural to assume that in a turbulent flow... the... fluid dynamic variables, will be random fields* (Monin and Yaglom 1971, p. 214), and studying from the outset, for example, probabilistic measures in some function space that are supported on Navier–Stokes solution, statistical solutions of the Navier–Stokes equations, etc. (Foiaş et al. 2001; Monin and Yaglom 1971; Vishik and Fursikov 1988). Another approach is replacing NSE by stochastic NSE, i.e. introducing a stochastic random force in the RHS of NSE both in theoretical physics, mathematics and computations with the belief that the SNSE with some "properly chosen" stochastic forcing are equivalent to some problems of pure NSE with numerous attempts to mimic the behavior of real turbulence in such a way, but there is hard evidence that different forcing of NSE—even in large scales only—result in different outcomes unless with the random force is properly chosen as in, e.g. RNG theories, see Biferale et al. (2004) and references therein, there is an issue of choice of not only random forcing in DNS. An obvious counter-example is the mentioned above stochastic forcing/excitation of integrable equations, e.g. Burgers, Korteveg de Vries, restricted Euler, etc., which without such forcing do not exhibit any "stochastic" behavior whatsoever. More generally random behavior of nonlinear systems as a response to random forcing is not necessarily turbulence and is an example among a multitude of qualitatively different phenomena—from Hamiltonian chaos to dissipative systems—all of which are qualified as "stochastic behavior".

We stress again that the statistical descriptions of the phenomenon of turbulence are not synonymous with the "statistical nature" of both the phenomenon and the problem. Both are different from each other and by no means are not exhausted by the term "statistical".

Another important distinction is between statistical theories and statistical methods of description and interpretation of the data from laboratory, field and numerical experiments. For example, in the former *ad hoc* assumptions are made in one way or another that a kind of low-dimensional description is possible with corresponding hypotheses mostly of unknown validity and obscured physical and mathematical justification about modeling/parametrization of the unresolved scales in terms of the resolved ones. As concerns the latter having fully resolved data from experiments and/or DNS one can get access to such key small scale quantities as the velocity derivatives (strain, vorticity, their production, etc.) providing the possibility of studying the rotational and dissipative nature of turbulence, along with other important quantities (e.g. fluid particles accelerations) for fundamental physics of the whole flow field rather than large or small scales only, keeping in mind that this flow decomposition is to a large extent artificial and in some sense even unphysical due to strong coupling between the two and due to the ambiguity of the very term 'scale' and the problematic nature of the decomposition approach to the phenomenon of turbulence and the necessity to handle turbulence as a whole.

5.2 Limitations of Statistical Methods

One of the problems of turbulent research is that we are forced to use statistical methods in one sense/way or another. All of them have inherent limitations the most acute reflected in the inability of all theoretical attempts to create a rigorous theory along with other inherent limitations of statistical methods, e.g. of handling data.

5.2.1 Statistical Theories

As mentioned this was recognized by many outstanding people. We just remind the two great probabilists Wiener (1938) and Kolmogorov (1985) who admitted that it is not clear how one has to treat rigorously turbulence using probabilistic approaches:

It has been realized since the beginning that the problem of turbulence is a statistical problem; that is a problem in which we study instead of the motion of a given system, the distribution of motions in a family of systems... It has not, however, been adequately realized just what has to be assumed in a statistical theory of turbulence (Wiener 1938).

From the very beginning it was clear that the theory of random functions of many variables (random fields), whose development only started at that time, must be the underlying mathematical technique... I soon understood that there was little hope of developing a pure, closed theory, and because of absence of such a theory the investigation must be based on hypotheses obtained on processing experimental data (Kolmogorov 1985), see Tikhomirov (1991, p. 487).

Note that both claim that *the problem of turbulence is a statistical problem* and both state that there is no such a theory in existence so that *the investigation must be based on hypotheses obtained on processing experimental data.*

The reason for the not optimistic second part of both statements above most probably is that as in other theoretical approaches the beautiful probabilistic instruments (probability as a measure, etc.) produced by Kolmogorov (1933) following the proposal by Borel (1909) appeared to be not instrumental for realization as concerns turbulence. Nevertheless, people in mathematical community insist that, e.g. such a theory should be based on the Navier–Stokes equations and their invariant measures and ergodic properties. Note the plural, which by itself is a problem as there exist unifinitely many and the choice is not a trivial matter as no one knows which one describes the real turbulence (Ruelle 1979 and references therein). Along with this problem there are several others. The general one is that is not all sure that a purely statistical rigorous approach if possible will be effective in solving the key physical problems—on the contrary, it seems that progress with the latter will aid the progress with the former.

The statements by Wiener and Kolmogorov were followed by the community in recognition that whatever the nature of the phenomenon and of the problem the intricacy and complex behavior of turbulence enforces use of statistical methods. This became a common view throughout the community and necessity also as concerns the theory of turbulence whatever this means, e.g. *The theory of turbulence by its very nature cannot be other than statistical, i.e., an individual description of the fields of velocity, pressure, temperature and other characteristics of turbulent flow is in principle impossible* and unavoidable *study of specific statistical laws, inherent in phenomena en masse,*[3] *i.e., in large ensembles of similar objects* (Monin and Yaglom 1971, p. 3).

---

[3]The *en masse* comes from the analogy with statistical physics. But there one has literally many similar objects—molecules. So one realization there may well suffice either, see below.

Just as in statistical physics, the technical reason that the statistical approach should be adopted from the outset in turbulence not only in 'theories', but also in handling the data from physical and numerical experiments. In both cases certain statistical hypotheses are made. But the former was quite successful in making a number of important predictions, whereas the latter, with few exceptions, such as the Kolmogorov four-fifths law (Kolmogorov 1941b) was unable to produce genuine predictions based on the first principles and did not result in any substantial progress of fundamental nature. All the rest—with all the respect to statistics—are postdictions and mimicking *after the experiments done* (Hoyle 1957). Apart from the above-mentioned reasons for such a failure it should be mentioned that, unlike in statistical physics, in turbulence neither 'simple objects'—such that a collection of these objects would adequately represent turbulent flows (perhaps the basic nonlinear objects of turbulence), if such exist and can be identified—'to do statistical mechanics' with them, nor 'right' statistical hypotheses have so far been found enabling. e.g. the "solution of the closure problem". The question about the very existence of both remains open. Nevertheless, major theoretical effort in statistical theories was made using various *ad hoc* assumptions of unknown validity and obscured physical and mathematical justification mainly on the relation of the small-scale structure with the rest of the flow. All this efforts are motivated by the natural tendency to simplify the problem which is manifested in numerous searches for a reduced/low dimensional description of turbulent flows. As mentioned the whole issue is closely related to the problem of decomposition/representation of turbulent flows. From the fundamental point it is not obvious that such a reduced adequate description does exist and/or is possible at all[4] because, e.g. it is missing essential physics contained in small scales associated with such fundamental properties of turbulence being a rotational and dissipative phenomenon, see Chaps. 7 and 8, and Chap. 3 in Tsinober (2009). Among the essential differences—apart of those mentioned in Monin and Yaglom (1971, pp. 4–5)—it should be mentioned that the small scales in turbulence (i) are very far from being simple as objects used in statistical mechanics and (ii) interact non-trivially, bidirectionally and non-locally, with the rest of the flow, i.e. contrary to the common view, the small scales cannot be seen as a kind passive sink of energy and/or as 'slaved' to the large scales—the small scales react back in a nonlocal manner and (iii) they carry lots of basic physics of turbulence. For this reason the following rather popular expectation appears to be conceptually incorrect: *It is both reasonable and realistic to expect that the removal of small scale and high-frequency components of the dynamics of a flow be described in terms of modified molecular transport coefficients like an eddy viscosity* (Orszag et al. 1993).

The eddy viscosity is one of the oldest and greatest analogies/misconceptions in the sense that it 'explains the enhanced transfer rates', etc. whereas it is just an empirical way of accounting for such rates, but not at all an explanation in any sense.

[4]*The basic question* (which usually is not asked) *concerning statistical description is whether such complex behavior permits a closed representation that is simple enough to be tractable and insightful but powerful enough to be faithful to the essential dynamics* (Kraichnan and Chen 1989).

The problem is that in such an approach the rotational and dissipative aspects are not considered as belonging to *the essential dynamics*.

Moreover, it is doubtful that any model except (hopefully) the NSE can be used to adequately study the physics of turbulent flows which in the first place means its basic/fundamental and conceptual aspects. *Perhaps the biggest fallacy about turbulence is that it can be reliably described (statistically) by a system of equations which is far easier to solve than the full time-dependent three-dimensional Navier–Stokes equations* (Bradshaw 1994).

In other words, the consequence of insisting on statistical theory as the only possibility in the theory of turbulence leads to the necessity of low-dimensional description with ***the removal of small scale and high-frequency components of the dynamics of a flow*** including quantities containing a great deal of fundamental physics of the whole flow field such as rotational and dissipative nature of turbulence among others. Thus, relying on statistical methods only (again with all the respect) one is inevitably loosing/missing essential aspects of basic physics of turbulence, see Chaps. 7 and 8.

The problem seems to be even a bit more complicated since turbulence—being studied by all kinds of statistical methods of description—cannot be considered as just a problem of statistical physics/mechanics only. There is no effective/satisfactory theoretical framework to handle turbulence—nothing new: this was stated by von Neumann (1949), though there are claims that turbulence can be seen also (but only in part!) as a problem of nonequilibrium statistical physics or whatever. For example, most of turbulent flows are only partly turbulent with the coexistence of regions with laminar and turbulent states in the same flow.

There are inherent limitations of any theory employing statistical methods as handling quantities like probabilities because probabilistic approaches avoid the details of time evolution (i.e. dynamics) and structure. Indeed, the fluid flow is described by a trajectory in phase space. It is obviously not described adequately only by the corresponding probability distributions and similar objects. Having the latter only, which is what one gets using statistical methods reflects, at least in part, our ignorance of the trajectories and the information associated with it such as dynamical and evolutionary aspects/details and structure(s) of turbulent flows, from relatively simple to the fine-scale structure of attractors (Guckenheimer 1986). In other words, statistical methods aim from the outset and allow to get only partial information, e.g. Monin and Yaglom (1971, pp. 8, 217). This is of special concern due to the troublesome question whether it is possible to penetrate into the fundamental physics of turbulence via statistics only. All this even if the aim as formulated by Orszag (1977) would be fully fulfilled: *Ideally, the objective of analytical turbulence theory is the exact calculation of all statistical properties of turbulence.*

There is an essential difference between the enforced necessity to employ statistical methods in view absence of other methods so far and the impossibility **in principle** to study turbulence via other approaches. This is especially discouraging all attempts to get into **more than just "en masse"**. Also such a standpoint means that there is not much to be expected as concerns the essence of turbulence using statistical methods.

Another important aspect is that use of statistical methods—together with excessive (over)generalizations such as functional analysis introducing objects with unknown relevance to real physical objects and unclear physical meaning—brings inherent problems of interpretation of experimental data. The issue of difficulties of interpretation of statistical objects introduced by humans is not new, see e.g. Orszag (1977). In particular, statistics only can be misused and misleading, so there is a difficult issue of interpretation, especially if the information is not in physical space, e.g. Fourier or any other decomposition. Here the question is also about what kind of statistics one has to use. It is directly related to the most difficult question on what are the problems, i.e., to the skill/art to ask the right and correctly posed questions, which is quite a problem in turbulence research.

It is for the above reasons that (meanwhile) our concern here is not with statistical theories of whatever nature as so far not useful in handling the fundamental issues: our concern is mostly much less ambitious—the focus is on statistical methods of description and interpretation of the data from laboratory, field and numerical experiments on turbulent flows via appropriate processing of the data. The latter is likely to be a prerequisite for any worthy 'theory' of turbulence.

5.2.2 Statistical Methods of Description and Interpretation of the Data from Laboratory, Field and Numerical Experiments

There are also limitations of statistical methods of description and interpretation of the data from laboratory, field and numerical experiments with each particular statistical tool having its own limitations. Well known examples are represented by relatively simple quantities like means, correlations, spectra, and probability density functions of various quantities. These tools like means and correlations smooth out some important qualitative features of (typical) individual realizations. The 'mean fields', e.g., large-scale averages of velocity or concentration of some species or particles, are smooth whereas the individual realizations are not. They are corrugated, highly intermittent and contain clusters/regions of high level of some quantity/ies (enstrophy, dissipation, passive tracer, reacting species, particles, etc.) surrounded by low level 'voids' of this quantity. In other words, 'standard' 'traditional' statistical methods to a large extent ignore the structure(s) of turbulent flows, which was the main reason for numerous objections against statistical methods often understood as averaging only. More subtle statistical properties of turbulent flows associated with their structure(s) both in small and large scales are important in many applications.

For instance, special information on small-scale structure(s) is needed in problems concerning, e.g., combustion, disperse multiphase flow, mixing, cavitation, turbulent flows with chemical reactions, some environmental problems, generation and propagation of sound and light in turbulent environments, and some special problems in blood flow related to such phenomena as hemolysis and thrombosis. In such problems, not only special statistical properties are of importance like those describing the behavior of smallest scales of turbulence, but also actual 'nonstatistical' features like maximal concentrations in such systems as an explosive gas which should be held below the ignition threshold, some species in chemical reactions, concentrations of a gas with strong dependence of its molecular weight on concentration (such as hydrogen fluoride used in various industries, e.g., in production of unleaded petrol) and toxic gases. Similarly, problems such as the manipulation (and possibly control) of turbulence and turbulence induced noise require information on large-scale structure(s) of turbulent flows far beyond such simple statistical characteristics as averages, correlations, spectra and PDFs.

One of the general claims is that *the only possibility in the theory of turbulence is statistical description, based on the study of specific statistical laws, inherent in phenomena en masse, i.e., in large ensembles of similar objects* and that *an individual description of the fields of velocity, pressure, temperature and other characteristics of turbulent flow is in principle impossible* (Monin and Yaglom 1971, p. 3). If so it is by itself a severe and pretty problematic limitation since in many cases one has to look not only at the properties of turbulent flows *en masse*, but also of special interest is prediction of the behavior of particular fields and specific properties of individual realizations like those involved in weather forecasting.

Investigators who prefer to look upon turbulence as a stochastic process may be interested in predicting the future statistical properties of developing or decaying turbulence, or simply in determining the statistical properties of stationary turbulence. At the same time they may have little interest in predicting future states of particular realizations. Indeed, it is likely to be some assumed basic unpredictability of individual fields of turbulent motion which has made the application of stochastic theory attractive to these investigators. There are nevertheless some (look for other examples with a single realization!!!) instances where prediction of the behavior of particular fields of turbulent motion is of considerable interest and importance. This is notably true in the case of weather forecasting. The atmosphere is, after all, a turbulent fluid; the migratory cyclones and anticyclones which bring us much of our weather are among the more conspicuous turbulent elements (Lorenz 1972, p. 195).

After all one does not need ensemble averaging to be sure that the coffee will be well mixed via only one and pretty short realization. Likewise, not much can be done statistical-wise to cope with a destructive hurricane or a tropical cyclone.

It may also be that such (i.e. very rare and exceptionally strong) events are rather sensitive to details of the physics that do not appreciably affect the character of the majority of events. This does not mean that one should not keep trying, by insight and discernment, to discover useful statistical measures, but rather that statistics will have to be used with that humility and appreciation of the combination of admission of ignorance and decision to ignore detail so successfully used by workers in the past (Mollo-Christensen 1973).

The problematic aspects of the claim that *the only possibility in the theory of turbulence* is the *study of specific statistical laws, inherent in phenomena en masse, i.e., in large ensembles of similar objects* are seen from the emergence of apparently random behavior in a system described by a purely deterministic system as NSE in **one** realization. In particular, some essential aspects of important physics of all these *similar objects* may be lost in handling them *en masse*. The issue is pretty well familiar to experimentalists who are using on a routine basis long enough single realizations to study various statistically stationary turbulent flows, or single snapshots of homogeneous flows in DNS on the basis of the ergodicity hypothesis, see below, which in a way eliminates the necessity to study the problem *en masse:* one "good" realization/or snapshot may be (and in many cases is) good enough. Figure 6.1 contains a clear indication how much can be lost as concerns time evolution looking just at the statistics.

5.2.3 On Particular Tools—Examples

Each particular statistical tool has its own limitations; being useful in one context/respect, it may say nothing in many others. A typical example is correlation, widely used in many aspects of turbulence research. Usually if a correlation between two quantities is not small, it reflects some important relation. However, if the correlation is small, it is not necessarily insignificant. For instance, let us have a look at the famous Reynolds stress $\langle u_1 u_2 \rangle$—the correlation between the velocity fluctuations in the direction of the mean flow (x_1) and those normal to the wall (x_2) in a wall-bounded turbulent flow. The typical value of the corresponding correlation coefficient is $\langle u_1 u_2 \rangle / u'_1 u'_2 \sim 0.4$. However, the real quantity entering the equation for the mean flow is the derivative $d\langle u_1 u_2 \rangle / dx_2$. In a developed turbulent flow with its mean properties independent of the streamwise coordinate, x_1 (flat channel, pipe), $d\langle u_1 u_2 \rangle / dx_2 = \langle (\boldsymbol{\omega} \times \mathbf{u})_1 \rangle \equiv \langle \omega_2 u_3 \rangle - \langle \omega_3 u_2 \rangle$. That is the 'turbulent force' is due to the coupling between large and small scales, i.e. nonlocality, see Chap. 7. The corresponding correlation coefficients between velocity and vorticity are small: both $\langle \omega_2 u_3 \rangle$ and $\langle \omega_3 u_2 \rangle$ are of order 10^{-2} or smaller. However, this does not mean that the coupling between $\boldsymbol{\omega}$ and $\mathbf{u}$ is insignificant. Indeed, without such a coupling $d\langle uv \rangle / dx_2 = 0$, so that the mean flow would not 'know' anything about turbulent fluctuations at all and therefore would remain as the laminar one. In this context a flow model with a mean homogeneous shear of infinite extent is pretty problematic. This is an example of more general problems with theoretical idealizations.

Moreover, even if a correlation between two quantities is very small or even precisely vanishing, this still does not necessarily mean that the interrelation/coupling between these two quantities is not existing or is unimportant. For example, in homogeneous turbulent flows, velocity and vorticity, and vorticity and the rate of strain tensor are precisely uncorrelated, $\langle \boldsymbol{\omega} \times \mathbf{u} \rangle \equiv 0$, $\langle \omega_i s_{ij} \rangle \equiv 0$, but their interaction is in the heart of the physics of any turbulent flow! Similarly, the correlation coefficient between $\mathbf{u}$ and $\nabla^2 \mathbf{u}$ is very small ($\sim Re^{-1/4}$) in high Reynolds number flows, but is very significant as directly related to the rate of dissipation of energy in turbulent flows. It is noteworthy that the above correlations are ≈ 0 in quasi (or locally)-homogeneous flows either.

One more example of the limited value of quantities like correlations and correlation coefficients is provided by a helically forced turbulent flow (Galanti and Tsinober 2006). In such a flow correlations between $\mathbf{u}$ and $\boldsymbol{\omega}$, and also $\boldsymbol{\omega}$ and curl $\boldsymbol{\omega}$ are not vanishing due to lack of reflectional symmetry. Nevertheless these correlations are an order of magnitude smaller than those between $\mathbf{u}$ and curl $\boldsymbol{\omega}$, and $\boldsymbol{\omega}$ and curl curl $\boldsymbol{\omega}$. This is in spite of the fact that the scales of $\mathbf{u}$ and $\boldsymbol{\omega}$ are 'closer' than those of $\mathbf{u}$ and curl $\boldsymbol{\omega}$ in the sense that the characteristic scales of $\mathbf{u}$ and $\boldsymbol{\omega}$ differ less that those of $\mathbf{u}$ and curl $\boldsymbol{\omega}$. Moreover, in flows with reflectional symmetry the correlation coefficients between $\mathbf{u}$ and $\boldsymbol{\omega}$, and $\boldsymbol{\omega}$ and curl $\boldsymbol{\omega}$ vanish, whereas correlations between $\mathbf{u}$ and curl $\boldsymbol{\omega}$, and $\boldsymbol{\omega}$ and curl curl $\boldsymbol{\omega}$ remain practically unchanged. The latter is directly related to the rate of dissipation of energy in turbulent flows as, e.g. in homogeneous flows $\langle \mathbf{u} \cdot \operatorname{curl} \boldsymbol{\omega} \rangle = -2\langle s_{ij} s_{ij} \rangle$.

Single point statistics in many cases may be and usually is insufficient and even misleading. For example, single point PDFs of velocity fluctuations are known to be quite close to the Gaussian distribution. In particular, the third moment of velocity fluctuations is close to zero, more precisely its skewness, $\langle u_1^3 \rangle / \langle u_1^2 \rangle^{3/2} \approx 0$, and the flatness, $\langle u_1^4 \rangle / \langle u_1^2 \rangle^2 \approx 3$, as in a Gaussian field. Similarly other higher order odd moments are small, and even moments assume values close to those of a Gaussian field, e.g. $\langle u_1^6 \rangle / \langle u_1^2 \rangle^3 \approx 15$. However, the conclusion that velocity fluctuations are really almost Gaussian is a misconception, not to mention the field of velocity derivatives. This is already seen when one looks at two-point statistics. For instance, in such a case the odd moments are significantly different from zero, e.g. Frenkiel et al. (1979). This is one of the simplest among numerous examples when multipoint in space and time statistics is useful. The widely known two-point correlations for some separation r and/or time t are related to the flow structure(s) larger than $\sim r/t$.

An example of special interest concerns the so called sweeping decorrelation hypothesis called also as random Taylor hypothesis:

Kolmogorov's basic assumption (Kolmogorov 1941a) *is essentially that the internal dynamics of the sufficiently fine-scale structure (in x-space) at high Reynolds numbers should be independent of the large-scale motion. The latter should, in effect, merely convect, bodily regions small compared to the macro scale* (Kraichnan 1959, p. 536). This is what is called sweeping which is claimed to have purely kinematic nature, and the claims that it preserves the shapes of the advected small scale eddies—this is why Lagrangian aspects are of importance—and thus has no effect on the turbulence energy spectrum in the Eulerian frame; *An underlying assumption of Kolmogorov theory is that very large spatial scales of motion convect very small scales without directly causing significant internal distortion of the small scales. The assumption usually is considered to be consistent with, and to imply, statistical independence of small and large scales* (Kraichnan 1964, p. 1723).

That is, this hypothesis is a reflection of a simplest heuristic decomposition of the flow field on large an small scales. The consequence is that, following Tennekes (1975), at large Reynolds numbers one can assume that *Taylor's "frozen-turbulence"*, random Taylor hypothesis—RTH, *approximation should be valid for the analysis of the consequences of large-scale advection of the turbulent microstructure and that the microstructure is statistically independent of the energy containing eddies.*[5] So that the fluid particle acceleration as assumed to vanish $\mathbf{a} = 0$

[5]The experimental evidence points to the opposite: the microstructure is not statistically independent of and even not decorrelated from the energy containing eddies, see below, Chaps. 7 and 8 below and Chap. 6 in Tsinober (2009) and references therein.

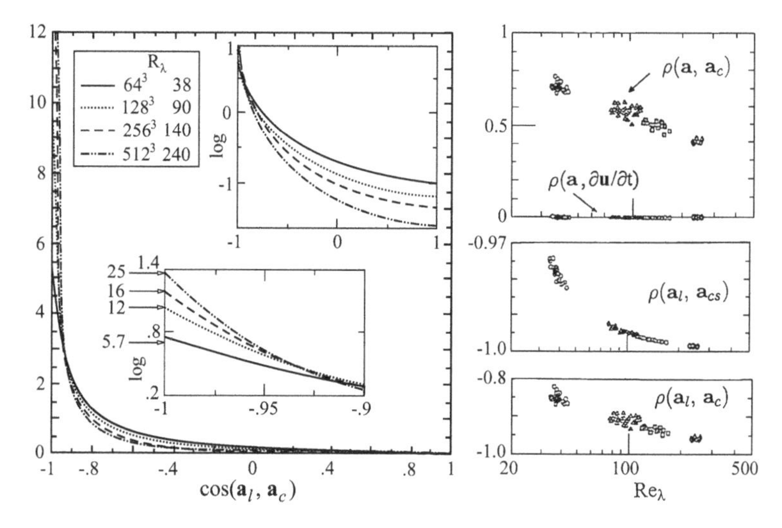

Fig. 5.1 *Left*: PDFs of the cosine of the angle between a_l and a_c. The *insets* show this dependence with the vertical in log and in the proximity of $\cos(\mathbf{a}_l, \mathbf{a}_c) \approx -1$. This alignment was observed first in DNS shown at the *left* (Tsinober et al. 2001), and also in laboratory experiments (Lüthi et al. 2005) and in the atmospheric surface layer (Gulitskii et al. 2007a, 2007b, 2007c). *Right*: Correlation coefficients between $\mathbf{a}_l$ and $\mathbf{a}_c$, and $\mathbf{a}_l$ and $\mathbf{a}_{cs}$, DNS (Tsinober 2001; Tsinober et al. 2001). The latter is the solenoidal part of $\mathbf{a}_c$. Similar results for $\mathbf{a}_l$ and $\mathbf{a}_c$ obtained in laboratory experiments (Lüthi et al. 2005) and in the atmospheric surface layer (Gulitskii et al. 2007b)

which is true if the sweeping is purely kinematic. On the other hand, the equality $\mathbf{a} = 0$ cannot and should not be understood literally because otherwise one is lead to absurd consequences. First, $\mathbf{a} = 0$ brings turbulence out of existence. Second, $\mathbf{a} = 0$ would mean that both $-\rho^{-1}\nabla p + \nu\nabla^2 u = 0$ and $\partial\mathbf{u}/\partial t + (\mathbf{u}\cdot\nabla)\mathbf{u} = 0$ are valid, which is wrong even as an "approximation". However, it is true that in some sense $\mathbf{a} \approx 0$, i.e. $\mathbf{a}$ is small compared both to the local acceleration, $\mathbf{a}_l = \partial\mathbf{u}/\partial t$ and $\mathbf{a}_c = (\mathbf{u}\cdot\nabla)\mathbf{u}$, e.g., $\langle a^2\rangle/\langle a_l^2\rangle \ll 1$ and $\langle a^2\rangle/\langle a_c^2\rangle \ll 1$. This in turn is possible if there is mutual (statistical) cancellation between $\mathbf{a}_l$ and $\mathbf{a}_c$. Since these quantities are vectors, the degree of this mutual cancellation should be studied both in terms of their magnitude and the geometry of vector alignments. Indeed, both in terms of magnitude, the total acceleration $\mathbf{a}$ is much smaller than its local and convective components $\mathbf{a}_l$ and $\mathbf{a}_c$, and that there is a strong anti-alignment between $\mathbf{a}_l$ and $\mathbf{a}_c$, see Fig. 5.1.

An important observation here is very large in magnitude values of correlation between $\mathbf{a}_l = \partial\mathbf{u}/\partial t$ and $\mathbf{a}_c = (\mathbf{u}\cdot\nabla)\mathbf{u}$, see Fig. 5.1b.

In other words the approximation $\mathbf{a} \approx 0$ is very good and becomes better with increasing Reynolds numbers. This is true also of the validity of the Random Taylor Hypothesis (RTH) or sweeping decorrelation hypothesis (SDH). The important point is that though both are *approximately* kinematic, this—as mentioned above—does not mean that the non-kinematic "small difference" justifies the validity of the equations $\partial\mathbf{u}/\partial t + (\mathbf{u}\cdot\nabla)\mathbf{u} = 0$ and $-\rho^{-1}\nabla p + \nu\nabla^2 u = 0$. *It is this "small non-kinematic difference" which is mostly responsible for all the dynamics in the*

Eulerian representation and that sweeping cannot be considered as just a kinematic effect. The dynamics involved is of utmost importance as there is no turbulence without the above mentioned "small difference"! This is because in the Kraichnan/Tennekes 'decomposition' there are two ingredients in the Eulerian decorrelation: (i) the sweeping of microstucture by the large scale motions and associated kinematic nonlocality, (ii) and the local straining, which is roughly pure Lagrangian. It appears that this kind of 'decomposition' is insufficient as it is missing an essential dynamical aspect—the interaction between the two. The random Taylor hypothesis, and, of course, the conventional Taylor hypothesis, lack/discard this aspect at the outset, which does not mean that these hypotheses are useless, but both are 'too kinematic'. It is noteworthy that the efforts on *elimination of the sweeping interactions from theories of hydrodynamic turbulence* are continuing (Gkioulekas 2007).

In the context of the discussion on correlations the above results show that very large correlations may be of little dynamical significance as contrasted to small counterparts of utmost importance: the decomposition on large and small scales appears "too kinematic" as it is missing an essential dynamical aspect—the direct and bidirectional interaction between the two.

In a more general context the above example shows that the statistical predominance not necessarily corresponds to the dynamical relevance as also seen from the recent example which concerns recent use of averaging of turbulent flow fields represented in the local coordinate system defined by the eigenvectors of the strain rate tensor at each point (Elsinga and Marusic 2010). Calculating the average fluctuating velocity field on a grid aligned with eigenframe of the strain revealed a shear layer at 45° with the most stretching and compressing eigendirections, which separates two larger-scale flow with regions of opposing velocity directions and vorticity associated with this shear layer preferentially oriented in the intermediate eigen-direction. The claim is that such *organization of the small-scale motions is not only found in the average patterns, but is also frequently observed in the instantaneous velocity fields of the different turbulent flows* and that *the presented average pattern in the strain rate eigenframe is representative of a general frequently occurring flow pattern* with similar claims throughout the paper. However, the problem is that all the evidence as presented by the authors (and some other) on the instantaneous fields is different from the above mentioned average, though with preferential alignment of vorticity and the intermediate eigen-direction. Moreover, though this alignment is statistically predominant it not most dynamically relevant as the flow patterns associated with the preferential alignment of vorticity and the largest positive eigen-direction contribute most to the enstrophy production. That is at best it is not clear in what sense the above mentioned average is *representative of a general frequently occurring* the instantaneous *flow pattern* if at all. Thus the approach of the authors is suspicious to what can be called oveprocessing of the data by "exotic" averaging.

Similar concern is arising in issues handling turbulence structure(s). These are finite objects which nevertheless are mostly hunted by isosurfacing and thresholding and are defined at some time moment only and, moreover, they cannot be followed in time. Therefore producing statistics is performed out of collections of "similar objects" obtained from snapshots at the same and different time moments. But the painful question is how really "similar" are all these if they are typically defined by one parameter? It is almost obvious that such kind of "statistical" processing, i.e. another kind of "exotic" averaging is killing most of essential features of the real "structure" and leaves the question of relevance, say the dynamical one, of these "structures" at best open.

The bottom line is that with all the respect, relying on statistics only such as, e.g. very strong and very weak correlations, 'exotic averaging', etc., may bring one to nowhere by missing essential dynamical effects. Judicious use of statistics is more than vital in turbulence research.

Chapter 6
Additional Issues of Importance Related to the Use of Statistical Methods

Abstract One of the concerns is that statistics only can be misused and misleading, so there are difficult issues of interpretation, validation and related, especially if the information is not in physical space, e.g. Fourier or any other decomposition.

One of the concerns is the issue of statistical dominance versus dynamical relevance. The statistical predominance not necessarily corresponds to the dynamical relevance as, e.g. in the case of sweeping decorrelation hypothesis or "exotic" averaging of turbulent flow fields such as represented in the local coordinate system defined by the eigenvectors of the strain rate tensor at each point.

Similar concern is arising in issues handling turbulence structure(s). These are finite objects which nevertheless are mostly hunted by isosurfacing and thresholding and are defined at some time moment only and, moreover, they cannot be followed in time. Therefore, producing statistics is performed out of collections of "similar objects" obtained from snapshots at the same and different time moments. But the painful question is how really "similar" are all these if they are typically defined by one parameter only? It is almost obvious that such kind of "statistical" processing, i.e. another kind of "exotic" averaging is killing most of essential features of the real "structure" and leaves the question of relevance, say the dynamical one, of these "structures" at best open.

The bottom line is that with all the respect, relying on statistics only such as, e.g. very strong and very weak correlations, 'exotic averaging', etc., may bring one to nowhere by missing essential dynamical effects.

Among the consequences is a serious misuse and interpretational (and terminological either) abuse of observations, which is essentially "aided" by the absence of genuine theory.

It is not unusual that when it goes about validation that the hard data (both from physical observations and numerical) is considered as kind of inferior as compared to 'models' so that the former are tested against the latter and not vice a versa as it is done even in fields being in possession of rigorous theories. Not many care about the observational evidence. The approach in reality is in some sense reverse. The widespread view of both mathematicians and theoretical physicists is that the main function of all experiments/observations both physical and numerical is to "validate theories"—paradoxically nonexistent so far. In such a situation the issue of interpretation and validation as concerns the right results for the right reasons

A. Tsinober, *The Essence of Turbulence as a Physical Phenomenon*,
DOI 10.1007/978-94-007-7180-2_6,

or "theories" versus hard evidence becomes of more than of utmost importance. A related issue of importance is on ergodicity and similar.

As mentioned statistical methods are practically the only employed in turbulence research be it theory (whatever this means) or experiment. Along with the high dimensional nature, very large number of strongly and nonlocally interacting degrees of freedom a number of issues take on special significance. Following is the list of those we consider as especially important for basic research in turbulence. Each are explained via short comments illustrated by way of exemplification.

6.1 Interpretation and Validation or What About the Right Results for the Right Reasons or Theories Versus Hard Evidence

The issues of interpretation and validation refer not just to "theories", but in the first place to the factual information and the relation between the two. Among the serious problems is that most of the "theories" claiming explanations of some specific aspects of observations in fact appear to be mere descriptions of these same particular observations or using these same observations for justifications/confirmations of these same theories.

6.1.1 Interpretation

This refers not just to "theories" but in the first place to experiments. The main point here is that the right results should be interpreted and related to the right reasons at least as concerns fundamental studies. The correspondence with experimental results may (and very frequently does—judging by massive agreement between "theories" on one hand and absence of genuine theory, on the other) occur for the wrong reasons,[1] i.e. this correspondence is at best a necessary condition. One of the problems is misinterpretation, which is "aided" by the extremely complex nature of the problem. Another acute problem is that most of the theories claim explanations of some specific aspects of observations being in fact mere descriptions of these same particular observations.

Before getting to specific examples we remind again the general issue of use of statistical methods together with excessive (over)generalizations such as functional analysis introducing objects with unknown relevance to real physical objects

[1] For example, *it is clear that if a result can be derived by dimensional analysis alone… then it can be derived by almost any theory, right or wrong, which is dimensionally-correct and uses the right variables* (Bradshaw 1994).

and unclear physical meaning which brings inherent problems of interpretation of experimental data.

The first example concerns the celebrated $k^{-5/3}$ energy spectrum which is considered as one of the basic attributes of many turbulent flows. In some sense this is correct, but there exist a multitude of phenomena (even not fluid dynamical, for more examples see Tsinober 2009, Chap. 7, p. 230, also pp. 211–216; Chap. 5, Sects. 5.3 and 5.4.5) possessing the same spectrum. Moreover, an extreme example is a single sharp change in velocity, which is local in physical space. Represented in Fourier space it has an energy spectrum $k^{-6/3}$ which is not so easy to distinguish from $k^{-5/3}$. This is true not only of the above particular scaling exponent,[2] but generally of scaling exponents as there exists no one-to-one relation between simple statistical manifestations and other more subtle properties such as the underlying structure(s) of turbulence, so that qualitatively different phenomena can and do possess the same set of scaling exponents. For example, there exist numerous models that attempted to reproduce the anomalous scaling for higher order statistics which are based on qualitatively different premises/assumptions, but all of which are in good agreement with the same experimental and numerical evidence, see Tsinober (2009, pp. 211–216) and references therein. Likewise similar PDFs of some quantities can correspond to qualitatively different processes, structure(s) and quantitatively different values of Reynolds numbers.

Another issue is the use of some theoretical considerations frequently called "theories" for idealized homogeneous and isotropic flows for "interpretation" of experiments which are neither homogeneous nor isotropic and even less: most of turbulent flows are only partly turbulent flows (PTF) with the coexistence of regions with laminar and turbulent states in the same flow, such as jets, wakes, mixing layers and some boundary layer flows.

In other words, there is a serious misuse of experimental data.

Two examples are given below.

The first example belongs to the category of spatially developing flows, self-similarity versus true time evolution. It concerns the use experimental data in partly turbulent flows and grid turbulence. Both are evolving in space statistically stationary *inhomogeneous* flows in spatially bounded domains with some boundary conditions. It should be stressed that the inflow conditions are indeed boundary conditions too—they are not initial conditions. Nevertheless, it is a common rather problematic practice to replace the evolution in space (in the streamwise coordinate) by the evolution in time. In particular, the grid turbulence data are frequently considered as decaying in *time* with the flow at each location x in the streamwise direction as representing an *imaginary* statistically spatially homogeneous turbulent flow in all

[2]It is noteworthy that this spectrum is not precisely the "right" one. Indeed, if one looks at the data by Grant et al. (1962), especially unpublished, but see Long (2003), the error bar is not that small as to exclude the $k^{-6/3}$ spectrum which correspond just to a single sharp change in velocity, see also Tsinober (2009, p. 334) and references therein for recent results on the "approximately" $k^{-5/3}$. Moreover, the "small" differences are essential and increase as concerns higher order quantities, derivatives and "strong events".

the three dimensions of infinite extent, which is obviously incorrect. Indeed, on one hand the $\partial/\partial t \neq 0$ for the imaginary three-dimensional state as it evolves, but not the x-dependence ($\partial/\partial x$) when looking at such a state at each time moment. The problematic aspect is not only with the streamwise direction, but also in the other two directions influenced by the boundaries. The "quasi-homogeneous" behavior of some quantities in bounded regions is misleading due to several factors, the main being the nonlocal nature of turbulence. From the conceptual point the nonlocality makes local homogeneity, isotropy, etc. impossible unless the whole (infinite extent) flow is such, which is trivially impossible.

A second example is the so called multifractal formalism (MF) (Frisch 1995), using the experimental data in turbulent flows of the kind mentioned above such as in the proximity of the centerline of a jet at moderate Reynolds numbers. The data is used to justify the MF and at the same time the MF is used to "explain" the data, see below.

The above extends into misinterpretations of analogies such as between the genuine (e.g. NSE) turbulence and passive "turbulence", i.e. evolution of passive objects in random (or just not too simple) velocity fields. The differences are more than essential, though there are numerous claims for *the well-established phenomenological parallels between the statistical description of mixing and fluid turbulence itself* (Shraiman and Siggia 2000), which are a consequence of multiple repetition of this claim without almost any factual basis. The first example is the famous verse by Richardson (1922, p. 6) related to the cascade picture of turbulent flows:

...we find that convectional motions are hindered by the formation of small eddies resembling those due to dynamical instability. Thus C.K.M. Douglas writing of observations from aeroplanes remarks: "The upward currents of large cumuli give rise to much turbulence within, below, and around the clouds, and the structure of the clouds is often very complex". One gets a similar impression when making a drawing of a rising cumulus from a fixed point; the details change before the sketch is completed. We realize thus that: big whirls have little whirls that feed on their velocity, and little whirls have lesser whirls and so on to viscosity—in the molecular sense.

The point is that this observation was made by looking at the structure of clouds, i.e., condensed water vapor, at the interface between laminar and turbulent flows in their bulk, which today is known not necessarily reflecting the structure of the underlying velocity field. The analogy between genuine and passive turbulence is illusive and mostly misleading, see Fig. 1.3, Chap. 1 and Tsinober (2009, Sects. 9.3, 9.4, 9.5.4) and references therein. In this context of special interest is a recent statement by Eyink and Frisch (2011, p. 362):

...Kraichnan's model of a passive scalar advected by a white-in-time Gaussian random velocity has become a paradigm for turbulence intermittency and anomalous scaling (the authors mean the genuine NSE turbulence)... The theory of passive scalar intermittency has not yet led to a similar successful theory of intermittency in Navier–Stokes turbulence. However, the Kraichnan model has raised the scientific level of discourse in the field by providing a nontrivial example of a multifractal field generated by turbulence dynamics. It is no longer debatable that anomalous scaling is possible for Navier–Stokes.

There are three points of problematic nature. First, the Kraichnan model it is not a paradigm for turbulence intermittency and anomalous scaling—it is even not a paradigm for intermittency and anomalous scaling for passive objects due to use of unphysical velocity field. Second, it is not an example of multifractal model, but rather an example of anomalous scaling which are definitely not synonymous.[3] And third, it is definitely not debatable that anomalous scaling is possible for Navier–Stokes, and even more the anomalous scaling **is in** the Navier–Stokes as it is observed in experiments for quite a while in real systems which are described by NSE. However, the real issue is about the reasons/interpretation or even better underlying mechanisms. The very problem is reflected by the term "anomalous scaling" as it is observed in what is called "inertial range" which in reality is an object in non-existence as not well defined due to severe contamination of the conventionally defined inertial range by strong dissipative events at whatever large Reynolds numbers, see Chap. 8. Far more can be found on misuse and misinterpretations of analogies in Chap. 9 (Tsinober 2009).

An additional example of interest is by Berdichevsky et al. (1996) in which mean velocity distributions were obtained '*from the first principles*' for *turbulent* Couette and Poiseuille flows, which are in very good agreement with experimental results for real three-dimensional flows. The problem is that these theoretical results are based on a two-dimensional model lacking any essential phenomena specific for three-dimensional flows such as vortex stretching whatsoever.

6.1.2 Validation or Theories Versus Hard Evidence

An intimately related to the just discussed issue of interpretation is the one concerning experimental validation of 'theories' understood as any theoretical treatment including modeling. This is directly related to the question on how meaningful and in what sense is the experimental 'confirmation' and/or 'validation' of a 'theory' as the first thing one is wondering about what one is supposed to validate and/or confirm. This is in the first place because in turbulence it is especially true that experimentalists observe what cannot be explained whereas theoreticians claim to explain these observations on the basis of idealizations which can not be realized. This statement is far more specific than it seems as we will see all along this text.

The highly dimensional nature of turbulence is one of the main reasons and obstacle for assessment of conceptual validity/reliability of any theory let alone low-dimensional (LD) modeling. From a conceptual point of view the main question remains whether it is at all possible and why does it 'work', of course, there is

[3]In this context it is of interest to quote Goto and Kraichnan himself (2004): *Multifractal models of turbulence have not been derived from the NS equation but they are supported by theoretical arguments and their parameters can be tuned to agree well with a variety of experimental measurements... Multifractal cascade models raise the general issue of distinction between what is descriptive of physical behavior and what can be used for analysis of data... Multifractal models may or may not express well the cascade physics at large but finite Reynolds numbers.*

a serious concern about the meaning of the term 'works'. Any LD model or some kind of a "theory" that represent a corresponding LD part/aspect of some particular kind/class of turbulent flows—but not necessarily for the right reason—will be (and usually is) inadequate in other flows. Just like simple interpolation/fits polynomials, etc. describe faithfully the behavior of data without any physical reason (purely empirical/technical), so many models do precisely the same. Mostly they are postdictions (rather than predictions) and, quite often, successful and useful semi-empirical interpolation schemes. There are many theories—many with contradictory premises—but all agreeing well with *some* experimental data. The issue is more serious as there are many situations in which agreement with experiment may not help too much even if the agreement between "theories" and experiment is excellent as the correspondence with the experimental results may occur for the wrong reasons as happens from time to time in the field of turbulence. For example, there are quantities/properties that are insensitive/invariant to some specific properties of the flow field whether it is real or in some sense synthetic, Gaussian/quasi-normal, Markovian, etc. For example, addressing issues associated with Gaussian/quasi-normal manifestations of turbulent flows and having a perfect agreement with some theory based on quasi-Gaussianity and/or quasi-normality an experimentalist encounters, in fact, a dilemma whether his measurements are perfect or just a nice Gaussian noise.

It is not unusual that when it goes about validation a frequent phenomenon is that the hard data (both from physical observations and numerical) is considered as kind of inferior as compared to 'models' so that the former are tested against the latter and not vice a versa as it is done even in fields being in possession of rigorous theories.

Returning to the mentioned above multifractal formalism (Frisch 1995), it is claimed to be an explanation of the 'anomalous scaling' and that the multifractal model is well supported by experimental evidence. In fact, it is another description of the observed anomalous scaling, i.e. of the experimental evidence employing at best some general properties such as basic symmetries and dimensional arguments, conservation laws and some other general properties etc. as the Navier–Stokes and even Euler equations. On top of this the experimental data used are all at best at moderate Reynolds numbers, whereas the MF is designed for $Re \to \infty$. As mentioned there are numerous alternative descriptions—sometimes standing in contradiction with each other—which are also "well supported" by the experimental evidence. A notable one is called breakdown coefficients/multipliers by Novikov (1990a).

It is should be stressed in all of the above that one of the key symmetries—the very existence of scaling exponents in the statistical sense, e.g., in the inertial range is assumed, i.e. it is a hypothesis only and is taken for granted, so it is a problem by itself. In other words such models are more or less successfully mimicking the experimental observations on the anomalous scaling, the multifractal formalism being (probably) the most successful as having the most freedom—a whole range of exponents and a function—claimed to be universal—for adjustment to the experimental data—at this stage the universality is forgotten. This issue (and some related) is even more serious as there are problems concerning the experimental evidence itself used.

We mention also the self-preservation theory of flows, e.g. past grids in the version as described in George (2012), see also references therein. Though it is termed as "theory" it is based on the spectral energy equation only, the spectral analogue of the von Karman and Howarth equation (1938), i.e. for low order statistics only. These equations contain more than one unknown, so that like von Karman and Howarth a "hidden closure" is introduced which in this case is the self-preservation hypothesis. Thus as any other model involving closure it cannot be considered as genuine theory. It is also noteworthy that this model (like the MF formalism) is in possession of considerable freedom for adjustment to the experimental data. There are many other attempts to "solve" some "equations of turbulence" by closing them using other "hidden closures", e.g. exploiting Lie group formalisms (Rosteck and Oberlack 2011; She et al. 2012; Hou et al. 2013) via multiscale analysis of the Reynolds stress in terms of the solutions of local periodic cell problems, see also references in the above papers.

An outstanding example of different nature is the Kolmogorov 4/5 law which is independent of and insensitive to the nature of dissipation mechanism as it depends on the mean energy injection rate only. Its validity at high Reynolds numbers in the form $S_3 \equiv \langle(\Delta u)^3\rangle = -4/5\langle\epsilon\rangle r$, i.e. with negligible viscous term $VT \equiv 6\nu d\langle(\Delta u)^2\rangle/dr$, which was interpreted in favor of existence of the inertial range. However, this common view is not acceptable, since the negligible viscous term VT in the Kolmogorov 4/5 law does not contain **all** the viscous contributions: the PDF of Δu and consequently S_3 contains a nonnegligible contribution from dissipative events which keep the 4/5 law precise—without the dissipative events just mentioned the 4/5 law does not hold! The bottom line is that the 4/5 law is not a purely inertial relation even at $Re_\lambda \approx 10^4$. We address this and related issues in Chap. 8. Two more examples are the Yaglom 4/3 law for the passive scalar and the Richardson pair diffusion law which are true for any random isotropic velocity field including the Gaussian ones.

It seems that validation of "theories" which are not based on NSE (or more generally some other first principles—if such exist) is not that meaningful from the fundamental point of view. The remaining are theories involving severe idealizations, which by themselves have to be tested far more carefully than it is usually done. These involve assumptions of statistical homogeneity and isotropy, and other symmetries and assumptions (such as self-similarity originated by von Karman and Howarth (1938) and subsequent different versions, see references in George (2012) and employing other "hidden symmetries", etc.) possessed by the original NSE with and/or without boundary conditions and other constraints, but not by statistical properties of flows in question. Hence the term "assumptions" in above.

The presence of boundaries prompted Kolmogorov (1941a) to postulate restoring in some statistical sense of all the symmetries locally in space/time[4] except for one involving scaling:

---

[4]Frisch (1995) presents this in the form of his hypothesis **H1** (p. 74), but omits to mention that it is due to Kolmogorov: there is no presentation of the hypothesis of local isotropy in his book.

It is noteworthy that Kolmogorov theory is in reality based on *similarity and dimensionality* and has no connection to NSE, see e.g. Monin and Yaglom (1971, p. 21): *The great attention paid*

...we think it rather likely that in an arbitrary turbulent flow with sufficiently large Reynolds number $Re = \frac{LU}{\nu}$ *the hypothesis of local isotropy is realized with good approximation in sufficiently small regions G of the four-dimensional space* (x_1, x_2, x_3, t) *not lying close to the boundaries of the flow or its other special regions.*

It is noteworthy that though assuming sufficiently large Reynolds numbers Kolmogorov did not assume anything about the scaling properties at the outset. In order to cope with this issue he introduced the concept of the inertial range (IR) for which he assumed independence of statistical flow properties of viscosity and postulated the validity of the scaling symmetry for statistical properties. This appeared to be in (approximate) agreement for the second order statistics, but not for orders equal or higher than four with increasing deviations with the order. It appeared that one of the main problems is the ill-posedness of the concept of the IR: it is postulated and widely believed that (ideally) there is no dissipation and viscosity does not play any role in IR. However, there is hard evidence that at least some properties of the conventionally defined IR do depend on the nature of dissipation at whatever large Re and that strong dissipative events, which live in the IR and which appear to be not so rare, make a nonnegligible impact on the behavior of traditionally inertial characteristics such as structure functions and are at the origin of the so called anomalous scaling The relation for the third order—the so called 4/5 law (Kolmogorov 1941b)—is a direct consequence of NSE, so that at high Reynolds numbers it is indeed confirmed experimentally. The experimental results for the second order and the validity of the 4/5 law enhanced the belief as concerned the higher order statistics, but not for long. As mentioned, recent experimental evidence shows the 4/5 law is not a purely inertial relation, see Chap. 8.

To put it simply, there is no such an object as an IR with properties independent of viscosity, so it its meaningless to look for models explaining (not just describing—there are great many of such) the intermittency in such a nonexistent object (i.e. IR), as the multifractal formalism is, for more see Chap. 8 and Tsinober (2009, Chap. 5, Sect. 5.3, pp. 102–110, also Chap. 7, p. 215).

Another important issue is the Reynolds number dependence. In the context of the issue of validation/interpretation it is disturbing that the experimental evidence was obtained at moderate Reynolds numbers for finite nontrivial statistically nonhomogeneous systems, including such flows as jets which are even only partially turbulent as they consist of coexisting turbulent-nonturbulent regions. Nevertheless, theoreticians claimed an 'explanation' of these same observations based on infinite statistically homogeneous objects/boxes and $Re \to \infty$. Today they say that these same effects of finite box are of special interest. Again one of the latest examples of

---

in this book to, similarity and dimensionality is also conditioned by the fact that Kolmogorov's theory of locally isotropic turbulence (which is based entirely on these methods) is given a great deal of space here. In other words, experimental validation of Kolmogorov (1941a) theory, as all theories of this kind, has a limited value. Again, *it is clear that if a result can be derived by dimensional analysis alone... then it can be derived by almost any theory, right or wrong, which is dimensionally-correct and uses the right variables* (Bradshaw 1994).

this kind is the "multifractal formalism". Another disturbing aspect of globally (and also locally) homogeneous turbulent flows is that in such flows, e.g. correlations of quantities which are in the heart of any turbulent flow like $\langle \boldsymbol{\omega} \times \mathbf{u} \rangle$ and/or $\langle (\mathbf{u} \cdot \nabla)\mathbf{u} \rangle$, $\langle \omega_j s_{ij} \rangle$ and/or $\langle (\boldsymbol{\omega} \cdot \nabla)\mathbf{u} \rangle$, $\langle s_{ik} s_{kj} \rangle$ and $\langle \partial^2 p / \partial x_i \partial x_j \rangle$ vanish. In particular, homogeneous turbulence with non-zero mean shear has constant Reynolds stresses with zero divergence, so that a field of homogeneous turbulence can have no effect on the field of mean velocity, if it stays homogeneous. In other words, generally, one cannot apply results for globally homogeneous turbulent flows to flows which are approximately homogeneous in a bounded region of flows, which are otherwise non-homogeneous due to effects of nonlocality. As mentioned the attraction of Kolmogorov hypotheses is that they concern the local properties of any turbulence, not necessarily homogeneous or isotropic or decaying or stationary, provided the Reynolds number is large enough. The consistency of these hypotheses is debated for quite a while both from the theoretical point of view, but as with other similar issues without much progress from the fundamental point (Gkioulekas 2007; Hill 2006) and references therein. Similar problems are encountered with anisotropy. One of the main problems is the inherent property of nonlocality questioning the validity of the above hypotheses on local homogeneity/isotropy in bounded flow domains surrounded by nonhomogenous/anisotropic flow regions, which is supported by experimental evidence, see Chaps. 7 and 8.

The "quasi-homogeneous" behavior of some quantities in bounded regions is misleading due to several factors, the main being the nonlocal nature of turbulence. From the conceptual point the nonlocality makes local homogeneity, isotropy, etc. impossible unless the whole (infinite extent) flow is such, which is trivially impossible.

A final note is that the idealized configurations such as global homogeneity, etc. are attractive for theoretical approaches, but they are useful for systems with dominating properties of local nature. This seems to be not the case with turbulence. Hence the problem with the utility of idealized configurations, not to mention that they were not so useful in view of absence of theory from first principles anyhow.

6.2 Ergodicity and Related

The general approach in statistical methods in theory is assumed to employ the concept of ensembles consisting of realizations of the "same" turbulent flow under nominally identical external conditions.[5] The problem is which probability should be assigned to each realization, so that the ensemble would correspond to most closely to physical reality and what is the relation of the statistics based on ensembles and statistics obtained from space-time information from a limited number of realizations because experiments cannot not usually be repeated a sufficiently large

[5] In the language of mathematicians *invariant probability measures*, and there is a question *which one is selected in experiments* (Ruelle 1983).

number of times for statistics large enough ensembles to be made. The ergodicity hypothesis is the tool which is used to cope with this problem.

For statistically stationary flows ergodicity is roughly equivalence of 'true' statistical properties, not only means/averages, but 'almost' all statistical properties, of an ensemble to those obtained using time series in one very long realization. A similar property is defined in space by replacing time by space coordinate(s) in which the flow domain has an infinite extension, at least in one direction. In other words, the ergodicity hypothesis in a way eliminates the necessity to study the problem*en masse:* one "good" realization/or snapshot may be and in many cases is good enough.

Though it is not known whether three-dimensional turbulent flows are ergodic,[6] it is common to use the ergodicity hypothesis in turbulence research, e.g., in physical and numerical experiments: turbulent flows are just believed to be ergodic at least the statistically stationary ones. In other words, in statistically stationary situations the time statistics obtained in experiments is believed to correspond to a unique probability measure invariant under time evolution. Also there seems to exist no direct evidence regarding the validity of the ergodicity hypothesis in turbulent flows except of one exploiting the property of a turbulent flow which is both statistically stationary in time and homogeneous in space (Galanti and Tsinober 2004). In such a flow its temporal and spatial statistical properties such as temporal statistics corresponding to a time series at a single point in space and spatial statistics based on a single time snapshot over the flow domain should be the same if the ergodic hypothesis is correct, see Fig. 6.1. One sees that indeed the temporal and spatial statistical properties shown in this figure are close to be identical. However, we repeat that though the statistics are the same there is far more one has to explore as concerns individual realizations—either one long time realization or a single time spatial snapshot. It should be emphasized that though this example is in favor of the ergodicity hypothesis, there are non-trivial problems such as outlined in Tsinober (2009, Sect. 3.7).

A tempting simplified interpretation or even better just illustration (no proofs, etc. so far) of the ergodicity property of statistically stationary flows is in terms of an attractor assumed to exist,[7] which is a very complex non-trivial object (Doering

---

[6]*The problem with this ergodicity assumption is that nobody has ever even come close to proving it for the Navier–Stokes equation* (Foiaş 1997), though some mathematical results, which are claimed to be relevant to turbulence are given in Foiaş et al. (2001). Namely, they have shown that there are measures—in the language of physics ensembles—on a function space that are time-invariant. However, invariance under time evolution is not enough to specify a unique measure which would describe turbulence. Another problem is that it is not clear how the objects that the authors have constructed and used in their proofs are relevant/related or even have anything to do with turbulence.

[7]Turbulent flows possess (empirically) stable statistical properties (SSP), not just averages but almost all statistical properties. In case of statistically stationary flows the existence of SSP seems to be an indication of the existence of what mathematicians call attractors. But matters are more complicated as many statistical properties of time-dependent in the statistical sense turbulent flows (possessing no attractor, but stable SSP) are quite similar at least qualitatively to those of statisti-

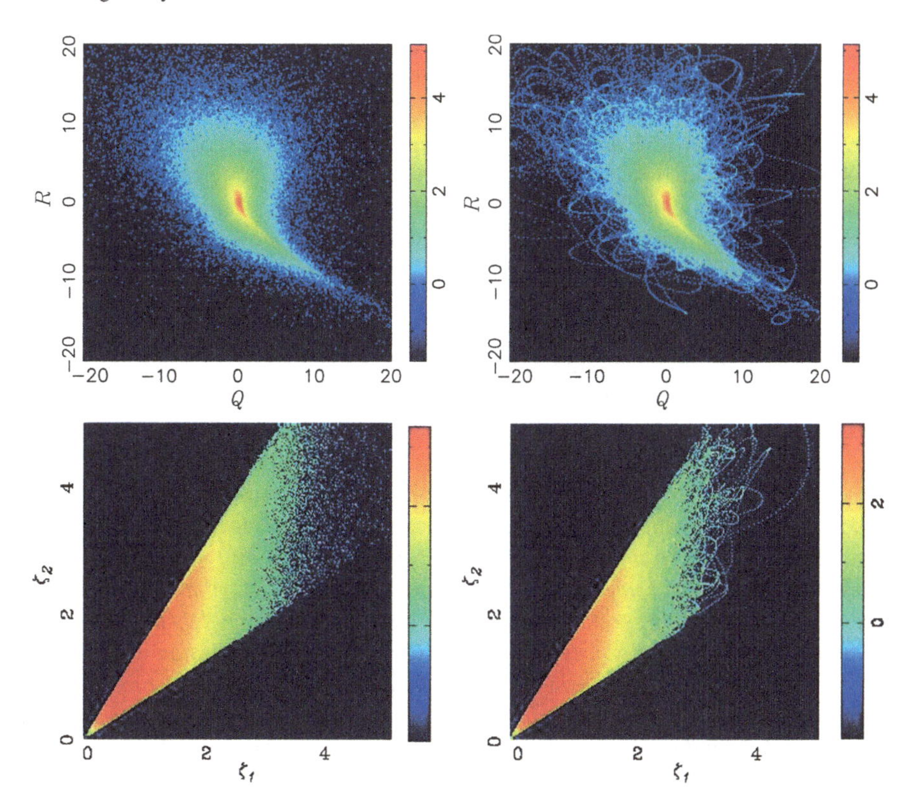

Fig. 6.1 *Top*: The 'tearing drop' pattern, which is the joint PDF of the invariants R, Q of the velocity gradient tensor $\partial u_i/\partial x_j$, $R = -1/3\{s_{ij}s_{jk}s_{ki} + (3/4)\omega_i\omega_j s_{ij}\}$, and $Q = (1/4)\{\omega^2 - 2s^2\}$. On the *left* is shown temporal statistics corresponding to a time series at single point in space. On the *right* is shown spatial statistics based on a single time snapshot over the flow domain. *Bottom*: An example of joint statistics of three quantities related by incompressibility relation—joint PDF of the eigenvalues of the rate of strain tensor Λ_1, Λ_2, Λ_3 in the plane $\Lambda_1 + \Lambda_2 + \Lambda_3 = 0$. Note that the time statistics in the *figure on the right* shows clear traces of time evolution, whereas nothing of the kind is observed with the spatial statistics on the left as it corresponds to one time snapshot and in this sense has nothing to do with the time evolution (Galanti and Tsinober 2004)

and Gibbon 2004; Foiaş et al. 2001; Robinson 2001). Almost any particular long enough flow realization comes very close to almost (i.e. with exception of measure zero) points of the attractor, and so will do almost any other (very long) realization. In other words, the solutions corresponding to the trajectories in the phase space will asymptotically visit (almost) all the attractor provided that the time span is long enough. In this sense almost all realizations will be almost the same. Hence the almost the same statistical properties—not just means, i.e. ergodicity. These re-

cally stationary ones as long as the Reynolds number of the former is not too small at any particular time moment of interest. This can be qualified as some manifestation of qualitative temporal universality/memory.

alizations will differ in their instantaneous, i.e. particular appearance not only due to (the popular) sensitive dependence on initial conditions, but also due to instabilities/*complexity of the behavior due to complex structure of the attractor* Arnold (1991) almost at any time and point of phase space and physical space either. Tritton (1988) defines turbulence as a *state of continuous instability*. Noteworthy is that the "same" above is essentially "statistically", their instantaneous "appearance" cannot be expected to be close in every respect. It is noteworthy that one cannot give a similar interpretation of spatially homogeneous flows as any single time snapshot does not reflect any time evolution, see again Fig. 6.1.

A related issue is predictability, e.g. in meteorology dealing with a more complicated issue ideally attempting prediction of individual realization in spite of the general claim that two initially nearly (but not precisely) identical turbulent flows become unrecognizably different on the time scale of dynamical interest. In reality, predictability in meteorology is dealing with prediction of (statistical) large scale properties of some "narrow sub-ensemble" of realizations based on models (like LES) with—again—eliminating the "small scales", but not totally: *The problem faced by anyone trying to model weather and climate is that we cannot totally ignore the unresolved scales of motion* (Palmer 2005). Instead, as in any modeling approach *we try to represent the unresolved scales in climate models by imagining an ensemble of sub-grid processes in approximate secular equilibrium with the resolved flow. The ensemble-mean (or "bulk") effect of these sub-grid processes is then given by a set of relatively simple (e.g. diffusive-like) deterministic formulae. We call such formulae "parametrizations" of the sub-grid processes* (Palmer 2005). One of the fascinating issues of fundamental nature is the dynamics of an initial error, which is the measure of the differences between some two realizations of a turbulent flow under almost (hence the error) the same conditions. An important aspect is that the error also possesses stable statistical properties in the sense that errors corresponding to different pairs of realizations have the same statistical properties e.g. see Tsinober and Galanti (2003).

Another related issue concerns reproducibility beyond just the overall statistical, which, in fact, to a large extent is the same as the one of predictability in meteorology. Far better predictions may be possible when a system is under strong influences like rotation, stable stratification (Cullen 2006) and similar; with high level of control of, e.g. inflow conditions as in a recent example in a boundary layer flow (Borodulin et al. 2011) and a far broader general theme of flow control including turbulent flows mainly in engineering (Gad-el Hak and Tsai 2006), and mathematical sciences (Barbu 2011) and references therein.

Part III
Issues of Paradigmatic Nature II: Specific Features

This part is a continuation of the previous part, which included a discussion of the consequences of complex behavior of systems described by purely deterministic equations including the necessity of change of the paradigmatic meaning of apparent randomness, stochasticity of turbulence which is roughly just the complexity due to a large number of strongly interacting degrees of freedom governed by the Navier Stokes equations.

It is devoted to more specific than in the previous part issues, but not less important. What follows comprises or are related to the second part of the major qualitative universal features of turbulent flows briefly described/listed in Chap. 1. These include the N's of turbulence with the emphasis on nonlinearity and nonlocality and the impact of the latter on such issues as large Reynolds number behavior and related. Here we make also a distinction between the issues of paradigmatic nature and those which are apparently/seemingly (pseudo-) paradigmatic such as the as the most popular cascades. A noteworthy feature is that the arguments in Chaps. 7 and 8 are supported by result from experiments at high Reynolds numbers not accessible to DNS of NSE with pointwise access to the full tensor of velocity derivatives, e.g. vorticity, strain among many others. Intermittency and structure(s) of and/in turbulence are overviewed in the last chapter taking into account the relevant aspects from the previous material.

Chapter 7
The N's of Turbulence

Abstract We start with the N's of turbulence. These comprise most of why turbulence is so impossibly difficult along with the essential constructive aspects facilitating all what is found in this book, i.e. to a large extent the "essence" as well. Whatever the approaches there are important common issues/difficulties/features most of which belong to this category: nonlinearity, nonlocality (and consequently "nondecomposability") and non-integrability, non-Gaussianity and non-Markovianity, nonequilibrium and (time) nonreversible, no scale invariance and no other symmetries, no small parameters and no low-dimensional description, and as a consequence of all this no theory based on first principles as NSE equations, which is a real frustration for a theoretician.

We concentrate on the most acute—to our view—with brief remarks on the rest, which does not mean that they are unimportant. These are in the first place the nonlinearity and nonlocality. Among a number of important roles of nonlinearity we stress also that it is responsible for the key properties of turbulence as an essentially rotational and strongly dissipative phenomenon. Nonlocality is a generic internal property of turbulent flows and exists independently of the presence of mean shear or other external factors, but has different and rich manifestations for different external factors. One of such manifestations is the direct and bidirectional coupling between large and small scales. We comment on the problematic issue of a variety of claims for locality which are not unrelated with the interpretational abuse of observations.

7.1 Non-integrability

In 1788 Lagrange wrote: *One owes to Euler the first general formulas for fluid motion... presented in the simple and luminous notation of partial differences... By this discovery, all fluid mechanics was reduced to a single point analysis, and if the equations involved were integrable, one could determine completely, in all cases the motion of a fluid moved by any forces* (Lagrange 1788, Sect. X, p. 271).

The **if** in the above citation is crucial: the Navier–Stokes equations are not integrable. Integrable systems, such as those having a solution 'in closed form' exhibit regular organized behavior, even those having an infinite number of strongly coupled degrees of freedom. A prominent example is provided by the solitons in the

A. Tsinober, *The Essence of Turbulence as a Physical Phenomenon*,
DOI 10.1007/978-94-007-7180-2_7,

systems described by the Korteveg–de Vries and Shrödinger equations. Two other examples are the Burgers and the so-called restricted Euler equation, which are integrable equations, and exhibit random behavior only under random forcing and or initial conditions, otherwise their solutions are not random.[1] That is, these examples represent the response of nonlinear systems to random forcing and which otherwise are not random, and should be distinguished from problems involving genuine turbulence. Navier–Stokes equations at sufficiently large Reynolds number have the property of intrinsic stochasticity in the sense that they possess mechanisms of self-"randomization" most probably at all scales, which are not fully understood. An important point is that non-integrability is intimately related to chaotic/complex behavior.

7.2 Nonlinearity

The role of nonlinearity—the general one, the chaotic/complex behavior of the fluid flow was discussed in Part I, it is also the most frequently fingered as the main 'guilty party' for the difficulties and for almost everything as concerns turbulence. This is definitely true as without nonlinearity there are no instabilities, bifurcations, transitions and turbulence itself. It is nonlinearity which makes life difficult due to massive use of decompositions. But, as we will see below, it is not alone. And there are several 'howevers'.

First, there are nonlinear problems that are completely integrable. The well-known examples, are systems displaying solitons or solitary waves. In these systems the many degrees of freedom are so strongly coupled that they do not display any chaotic/irregular behavior. Instead they are entirely organized and regular, see Zakharov (1990), Kosmann-Schwarzbach et al. (2004) and references therein. By a quite questionable and pretty problematic analogy it is thought for quite a while without almost any constructive output that the so-called coherent structures in turbulent flows may be viewed and treated viewed in a similar way, e.g. (Newton and Aref 2003).

Second, nonlinearity is frequently blamed for the difficulties in the closure problem which is associated with some form of decomposition, such as the Reynolds decomposition of the flow field into the mean and the fluctuations, or similar decompositions into resolved and unresolved scales associated with large eddy simulations.

[1]There is no consensus on the meaning of the term integrability, but it is agreed mostly that integrable systems behave nicely and are globally 'regular', whereas the nonintegrable systems are not 'solvable exactly' and exhibit chaotic behavior, see Zakharov (1990) and Kosmann-Schwarzbach et al. (2004) for more examples and discussion on what is integrability. The latter write *It would fit for a course entitled "Integrability" to start with a definition of this notion. Alas, this is not possible. There exists a profusion of definitions and where you have two scientists you have (at least) three different definitions of integrability but mention the definition by Poincar´e: to integrate a differential equation is to find for the general solution a finite expression, possibly multivalued, in a finite number of functions.*

The essence of the problem is that the equations for the mean field (resolved scales) contain moments of the fluctuations (unresolved scales) due to the nonlinearity of the NSE. However, a similar problem exists for the so-called advection-diffusion equation describing the behavior of a passive scalar in some flow field. But this equation is linear. The problem arises due to the multiplicative nature of the velocity field, since velocity enters this equation as its coefficients. Finally, a noteworthy caveat is that the inertial nonlinearity in the Euler setting, $(\mathbf{u} \cdot \nabla)\mathbf{u}$ have a relative nature as depending on the frame of reference—it is not Galilean invariant and in pure Lagrangian setting is even not present—the acceleration in pure Lagrangian setting is linear so that the inertial nonlinearity consists only of a term with pressure. Getting back to Eulerian setting it is noteworthy that the nonlinearity $(\mathbf{u} \cdot \nabla)\mathbf{u}$ consists of two parts, irrotational and solenoidal, $(\mathbf{u} \cdot \nabla)\mathbf{u} = \nabla(\alpha + u^2/2) + \nabla \times \boldsymbol{\beta}$, where $\nabla\alpha$ and $\nabla \times \boldsymbol{\beta}$ are the irrotational and solenoidal parts of the Lamb vector $\boldsymbol{\omega} \times \mathbf{u} = \nabla\alpha + \nabla \times \boldsymbol{\beta}$. We remind also that there is strong cancellation between $\partial\mathbf{u}/\partial t$ and $(\mathbf{u} \cdot \nabla)\mathbf{u}$ and so that the irrotational part of the sum $\mathbf{a} = \partial\mathbf{u}/\partial t + (\mathbf{u} \cdot \nabla)\mathbf{u}$, i.e. the acceleration is dominated by pressure, but in the norm $\langle(\cdots)^2\rangle$, see DNS by Vedula and Yeung (1999). However, it is the rotational part of acceleration $\mathbf{a}_s = \partial\mathbf{u}/\partial t + \nabla \times \boldsymbol{\beta} = \nu\nabla^2 u$ without which there is no turbulence.

7.2.1 Nonlinearity Plus Decompositions Gives Birth to "Cascades"

Cascade is essentially a phenomenological creature. Though phenomenology is not a genuinely fundamental problem/issue, see Chap. 5 in Tsinober (2009), there are several aspects of paradigmatic nature. Hence a short digression. First, as concerns decompositions any decomposition results in a nontrivial bidirectional relation between small and the large scales (whatever this means) which is non-local (functional) both in space and time, i.e. history-dependent. Nonlinearity along with any decomposition/representation of a turbulent field results in a process of interaction/exchange of (not necessarily only) energy between the components of the decomposition and, in particular, in a "cascade" with its properties depending on the "physics" of the specific decomposition.

This alone plus nonlocality (see below) makes the popular concept of cascade ill posed which along with other important issues calls for reminding the acute and generic problems arising from employing decompositions. Indeed, one of the problems with decompositions is that the nonlinear term redistributes, e.g. the energy among the components of a particular decomposition in a different way for different decompositions, i.e. the energy exchange/transfer is decomposition dependent, though one would expect that energy transfer, just like any physical process, should be invariant of particular decompositions/representations of a turbulent field. On the contrary the production of strain, i.e. dissipation which is the final destination of the assumed cascade does not care about decompositions. In other words, the energy 'cascade' (whatever this means, if anything) is associated primarily with the quantity $-s_{ij}s_{jk}s_{ki}$ responsible for the strain production, rather than with the enstrophy production $\omega_i \omega_j s_{ij}$ and that vortex stretching suppresses the cascade and does not aid it, at least in a direct manner. Consequently, it is the vortex compression, i.e. $\omega_i \omega_j s_{ij} < 0$, that aids the production of strain/dissipation and, in this sense, the 'cascade', see below. So there is a problem with *the classical energy cascade picture, in which vortices of a given scale are stretched by and absorb energy from structures of a somewhat larger scale* (Leung et al. 2012 and references therein). It is noteworthy that this and similar *classical views of the energy cascade* are based mainly on an erroneous analogy with material lines and other passive vectors, see Sect. 9.4 Vorticity versus passive vectors in Chap. 9 on analogies and misconceptions in Tsinober (2009).

The notion that turbulent flows are hierarchical, which underlies the concept of the cascade, though convenient, is more a reflection of the unavoidable (due to the nonlinear nature of the problem) hierarchical structure of models of turbulence and/or decompositions rather than reality, see e.g. the rigorous treatment of cascade between the components based on eigensolutions of the Stokes operator in Foiaş et al. (2001).

The class of flows called partly turbulent comprises a strong counter example to the cascade. In all such flows the fluid becomes turbulent in 'no time' without any cascade whatsoever. The ill-posedness of the cascade concept is emphasized in the case of passive objects, whose evolution is governed by linear equations, with the velocity field entering multiplicatively in these equations, thus making them 'statistically nonlinear'. In the Lagrangian description the inertial effects are manifested only by the term containing pressure. Therefore, the nonlinearity in the Lagrangian representation cannot be interpreted in terms of some cascade. An extreme example is the absence of a "cascade" in Lagrangian chaotic, but Eulerian laminar flows. There is much confusion of fluxes in inhomogeneous flows with cascades, e.g. the recent example by Jimenez (2012).

We should also mention two examples from stability related to the above. The first is about the of spatially periodic flows, which may destabilize directly into small-scale three-dimensional structures (Pierrehumbert and Widnall 1982). The second example shows that significant variations down to very small scale can be produced by a single instability at much larger scale without any 'cascade' of successive instabilities (Ott 1999).

The bottom line is that the cascade picture of turbulence is more a reflection of the hierarchic structure of various models of turbulent flows rather than reality. Most of these models have no connection with Navier–Stokes equations. Other relay severely on closures.

7.2.2 Turbulence Is Essentially Rotational and Strongly Dissipative Phenomenon

It is the Eulerian setting which enables one to observe the following key paradigmatic properties of turbulence associated with the nonlinearity of NSE.

There are two concomitant qualitatively universal physical mechanisms turning turbulence into a strongly dissipative and rotational phenomenon. These are the predominant production of the rate of strain tensor, s_{ij} and vorticity, ω_i. The rate of strain tensor, s_{ij} and vorticity, ω_i are just the symmetric and antisymmetric parts of the tensor of velocity derivatives $A_{ij} = \partial u_i/\partial x_j \equiv s_{ij} + \frac{1}{2}\varepsilon_{ijk}\omega_k$. Though formally the two representations are trivially equivalent, it is also trivially obvious that it is more appropriate from the physical point of view to use strain and vorticity as corresponding to the two fundamental properties of turbulence as strongly dissipative and essentially rotational. We emphasize that the direct causal relation to dissipation is not the only role played by strain $\epsilon = 2\nu s_{ij}s_{ij}$ in turbulent flows, rather than to enstrophy.

The mechanisms of strain and vorticity production are reflected by the terms $-s_{ik}s_{kj}$ and $\omega_j s_{ij}$ in the equations for s_{ij} and ω_i correspondingly. It should be emphasized that these terms are Galilean invariant unlike the term $(\mathbf{u}\cdot\nabla)\mathbf{u}$ they take their origin is not Galilean invariant. Likewise the terms $-s_{ij}s_{jk}s_{ki}$ and $\omega_i\omega_j s_{ij}$ responsible for the production of the total strain $s^2 \equiv s_{ij}s_{ij}$ and the enstrophy ω^2 correspondingly. These are the outstanding quantities of third order and the empirical fact of paradigmatic importance is that both are positively skewed, so that both $-\langle s_{ij}s_{jk}s_{ki}\rangle$ and $\langle\omega_i\omega_j s_{ij}\rangle$ are positive.

In contrast to the common view: *It seems that the stretching of vortex filaments must be regarded as the principal mechanical cause of the high rate of dissipation which is associated with turbulent motion* (Taylor 1938a, 1938b) it is the production of strain which is responsible both for (i) the enhanced dissipation of turbulence and in particular, for what is called "cascade" as resulting in enhanced dissipation, which is not surprising as the appropriate level of dissipation moderating the growth of turbulent energy is achieved by the build up of strain of sufficient magnitude and (ii) the enstrophy production either. In other words, apart of dissipation the strain field plays the role (among several others) of an engine producing the whole field of velocity derivatives, both itself and the vorticity, with compression aiding the prevalent production of strain and stretching aiding the prevalent production of enstrophy. It is of special importance on paradigmatic level that it is the strain production which is responsible for the finite overall dissipation at (presumably) any however large Reynolds numbers in contrast to two-dimensional flows where s^2 is an inviscid invariant.

The fascinating aspect of the above non-conformistic statements is that it becomes literally obvious when one takes the labor to look at both equations, i.e. for ω^2 and for s^2 too.

$$\frac{1}{2}\frac{D\omega^2}{Dt} = \omega_i\omega_j s_{ij} + \nu\omega_i\nabla^2\omega_i + \varepsilon_{ijk}\omega_i\frac{\partial F_k}{\partial x_j},$$

$$\frac{1}{2}\frac{Ds^2}{Dt} = -s_{ij}s_{jk}s_{ki} - \frac{1}{4}\omega_i\omega_j s_{ij} - s_{ij}\frac{\partial^2 p}{\partial x_i \partial x_j} + \nu s_{ij}\nabla^2 s_{ij} + s_{ij}F_{ij}.$$

It is seen also that the enstrophy production $\omega_i\omega_j s_{ij}$ appears in RHS of the equation for s^2 with the negative sign, so that the vortex stretching is opposing the production of dissipation/strain: all instantaneous positive values of $\omega_i\omega_j s_{ij}$ make a negative contribution to the RHS in the equation for s^2, i.e. enstrophy production $\omega_i\omega_j s_{ij}$ has an additional role as drain of "energy" of strain, i.e. s^2.

Since $\omega_i\omega_j s_{ij}$ is essentially a positively skewed quantity, its mean contribution to the strain production is negative. In other words, the energy 'cascade' (whatever this means, if anything) is associated primarily with the quantity $-s_{ij}s_{jk}s_{ki}$, rather than with the enstrophy production $\omega_i\omega_j s_{ij}$ and that vortex stretching suppresses the cascade and does not aid it, at least in a direct manner. Consequently, it is the vortex compression, i.e. $\omega_i\omega_j s_{ij} < 0$, that aids the production of strain/dissipation and, in this sense, the 'cascade', for more see Tsinober (2009, Sect. 6.2.2, pp. 128–133); also Schumacher et al. (2011).

The term $s_{ij}\Pi_{ij}$, with $\Pi_{ij} \equiv \partial^2 p/\partial x_i \partial x_j$, is a divergence, so it is vanishing in the mean in "locally"-homogeneous flows. However, $s_{ij}\Pi_{ij}$ is still of importance as, e.g. the PDF of $s_{ij}\Pi_{ij}$ is asymmetric and is positively skewed at large s^2 (Tsinober 2000). The term $s_{ij}\Pi_{ij}$ becomes important in inhomogeneous flows especially in purely irrotational inhomogeneous ones, e.g. in the nonlocal production of strain on the laminar side (not only in the proximity) of the turbulent non-turbulent "interface" in partly turbulent flows. It is noteworthy that the evolution equations for $\omega_i\omega_j s_{ij}$ and $s_{ij}s_{jk}s_{ki}$ contain important nonlocal terms $\omega_i\omega_j \frac{\partial^2 p}{\partial x_i \partial x_j}$ and $s_{ik}s_{kj} \frac{\partial^2 p}{\partial x_i \partial x_j}$, see Appendix C in Tsinober (2009).

An important a bit subtler aspect is that the field of strain is efficient in the above two missions only with the aid of vorticity, i.e. only if the flow is rotational, since otherwise the strain (self-)production, $s_{ij}s_{jk}s_{ki}$, for an irrotational flow field is just a divergence $s_{ij}s_{jk}s_{ki} = \partial\{\cdots\}/\partial x_i$. In this context the vortex stretching is necessary to support the rotational nature of turbulence. Otherwise. e.g. it is the term $s_{ij}\frac{\partial^2 p}{\partial x_i \partial x_j}$ which is mainly responsible for the strain production on the laminar side of the turbulent non-turbulent "interface" in partly turbulent flows, see the upper part of Fig. 1.1. The above does not exhaust the roles of strain, see Sect. 6.22 in Tsinober (2009).

There is a conceptual and qualitative difference between the nonlinear interaction between vorticity and strain, e.g. $\omega_i\omega_j s_{ij}$ and the self-amplification of the field of strain, $-\ s_{ij}s_{jk}s_{ki}$, which is a specific feature of the dynamics of turbulence having no counterpart (more precisely analogous—not more) in the behavior of passive and also active objects. This process (i.e., $s_{ij}s_{jk}s_{ki}$) is local in contrast to $\omega_i\omega_j s_{ij}$, as the field of vorticity and strain are related nonlocally.

Unlike the strain self-production the interaction of vorticity and strain involve important issues of geometrical nature which are complicated by the nonlocal relation between them. For example, $\omega_i\omega_j s_{ij} = \omega^2 \Lambda_k \cos^2(\boldsymbol{\omega}, \boldsymbol{\lambda}_k)$ so that the enstrophy production is essentially dependent on (i) the magnitude of ω^2, (ii) the eigenvalues Λ_i of the rate of strain tensor s_{ij} ($\Lambda_1 > \Lambda_2 > \Lambda_3$; due to incompressibility $\Lambda_1 + \Lambda_2 + \Lambda_3 = 0$ so that $\Lambda_1 > 0$, $\Lambda_3 < 0$), (iii) the alignments between vorticity $\boldsymbol{\omega}$ and the eigenframe $\boldsymbol{\lambda}_i$ of the rate of strain tensor s_{ij} and correlations between the three (i)–(iii). It appears that the main contribution to the enstrophy production and its rate is due to the first term associated with the $\boldsymbol{\omega}, \boldsymbol{\lambda}_1$

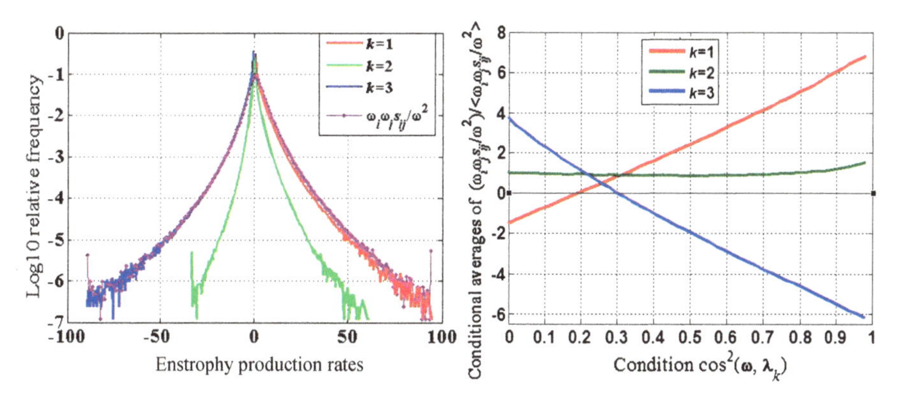

Fig. 7.1 (**a**) Histograms of the total rate of enstrophy production $\omega_i\omega_j s_{ij}/\omega^2$ and separate contributions $\Lambda_k \cos^2(\boldsymbol{\omega}, \boldsymbol{\lambda}_k)$, $k = 1, 2, 3$. It seen clearly that the main contribution to the total on the positive part comes from $\Lambda_1 \cos^2(\boldsymbol{\omega}, \boldsymbol{\lambda}_1)$. (**b**) Conditional averages of $EPR = \omega_i\omega_j s_{ij}/\omega^2$ on $\cos^2(\boldsymbol{\omega}, \boldsymbol{\lambda}_k)$. We preferred to use the evidence as obtained for real physical fields to avoid any abuse by some additional processing such as decompositions, etc. We stress that this is qualitatively different from the $\boldsymbol{\omega}, \boldsymbol{\lambda}_1$ alignment for some "parts" of the flows processed in some way, e.g. removing some 'local' part of strain or using band pass filtering, see Kevlahan and Hunt (1997), Porter et al. (1998), Hamlington et al. (2008), Leung et al. (2012)

alignment, see Tsinober (2009 and references therein). For example, in the field experiments with $Re_\lambda \approx 10^4$ the relation between the mean of the three contributions $\langle\omega^2\Lambda_1 \cos^2(\boldsymbol{\omega}, \boldsymbol{\lambda}_1)\rangle : \langle\omega^2\Lambda_2 \cos^2(\boldsymbol{\omega}, \boldsymbol{\lambda}_2)\rangle : \langle\omega^2\Lambda_3 \cos^2(\boldsymbol{\omega}, \boldsymbol{\lambda}_3)\rangle = 3.1 : 1.0 : -2.1$. The dynamical dominance of the term associated with the $\boldsymbol{\omega}, \boldsymbol{\lambda}_1$ alignment is much stronger for the corresponding rates, i.e. $\omega_i\omega_j s_{ij}/\omega^2 = \Lambda_k \cos^2(\boldsymbol{\omega}, \boldsymbol{\lambda}_k)$; $\langle\Lambda_1 \cos^2(\boldsymbol{\omega}, \boldsymbol{\lambda}_1)\rangle : \langle\Lambda_2 \cos^2(\boldsymbol{\omega}, \boldsymbol{\lambda}_2)\rangle : \langle\Lambda_3 \cos^2(\boldsymbol{\omega}, \boldsymbol{\lambda}_3)\rangle = 4.9 : 1.0 : -3.8$, which exhibits far stronger role of strain and $\boldsymbol{\omega}, \boldsymbol{\lambda}_i$ alignments. For PDFs of involved quantities see Fig. 7.1.

Thus it appears that the flow patterns with $\boldsymbol{\omega}, \boldsymbol{\lambda}_1$ alignment are dominant in enstrophy production in spite of the observed statistical predominance of the $\boldsymbol{\omega}, \boldsymbol{\lambda}_2$ alignment. In other words, the predominant vortex stretching is indeed due to alignment $\boldsymbol{\omega}, \boldsymbol{\lambda}_1$, but for this there is no need for the statistical predominance of this alignment as massively expected, since statistical dominance is not synonymous to dynamical relevance.[2] This apparent "contradiction" is resolved by noting that: (i) the intermediate eigenvalue, Λ_2, assumes both positive and negative values thus reducing the terms $\langle\omega^2\Lambda_2 \cos^2(\boldsymbol{\omega}, \boldsymbol{\lambda}_2)\rangle$ and $\langle\Lambda_2 \cos^2(\boldsymbol{\omega}, \boldsymbol{\lambda}_2)\rangle$ which involve both stretching and slightly less compressing; whereas Λ_1 is positive; and (ii) the magni-

[2]This expectation takes its origin mainly from the incorrect analogy with material lines and other passive vectors, see Sect. 9.4 Vorticity versus passive vectors in Chap. 9 on analogies and misconceptions in Tsinober (2009). In particular, the nonlocality of the relation between vorticity and strain—which does not have an analogue with passive vectors—play an important role in the issue of $\boldsymbol{\omega}, \boldsymbol{\lambda}_i$ alignments (Hamlington et al. 2008).

An important general aspect is that the strongest interaction between vorticity and strain occurs in regions with $\boldsymbol{\omega}, \boldsymbol{\lambda}_1$ alignments and large strain, see Chap. 6, pp. 150–153 in Tsinober (2009).

tude of Λ_1 is much larger, e.g. $\langle\Lambda_1^2\rangle : \langle\Lambda_2^2\rangle : \langle\Lambda_3^2\rangle = 10.2 : 1.0 : 13.7$, see Table 6.6 and Fig. 6.7 in Tsinober (2009).

We would like to recapitulate the qualitative differences between the enstrophy and strain production. It is the strain production (rather than vortex stretching) that is directly responsible for the enhanced dissipation of turbulent flows and it is a local process with predominant compressing whereas the enstrophy production is a nonlocal process with predominant stretching. To stress, though vorticity is commonly given more stress whereas in reality strain and vorticity are almost "equal partners". Almost because it is the strain production that plays the role of an engine producing the whole filed of velocity derivatives provided that the flow field is not vorticity-free.

The bottom line here is that on the paradigmatic level it is the nonlinearity that is responsible for such most basic key properties of turbulence as essentially rotational and strongly dissipative phenomenon. This is just because the excitation of small scales is due to the nonlinearity of the NSE.

It is worth of emphasizing that these two concomitant key properties and processes are observed in a rather straightforward manner, so that there is no need for "cascades", decompositions etc. These latter so far are mainly obscuring rather than helping to "understand the physics of cascades" or whatever, not to say about the positively skewed nature of the $-s_{ij}s_{jk}s_{ki}$ and $\omega_i\omega_j s_{ij}$, which is in the heart of the physics of turbulence. These quantities are among typical representing genuinely nonlinear processes in turbulence making it not amenable to quasi-nonlinear approaches such as RDT.

A final remark is that the dominance of strain and enstrophy production, $-s_{ij}s_{jk}s_{ki}$ and $\omega_i\omega_j s_{ij}$, as manifested, e.g. in the Tennekes and Lumley balance, see below, is challenged in regions lying close to the flow boundaries and other special regions such as those with large shear and laminar/turbulent interfaces in partly turbulent flows.

7.3 Nonlocality

7.3.1 Introductory/General Remarks

Turbulence is a nonlocal process. One can see clear hints to this from the term $(\mathbf{u}\cdot\nabla)\mathbf{u} \equiv \boldsymbol{\omega}\times\mathbf{u} + \nabla(u^2/2)$ and terms appearing in a number of precise consequences of NSE such as $(\boldsymbol{\omega}\cdot\nabla)\mathbf{u} \equiv \partial/\partial x_j\{u_i\omega_j\}$—so it is naturally to call $u_i\omega_j$ as helicity tensor, $\nabla^2 p = \rho(\omega^2 - 2s_{ij}s_{ij}) = -\rho\frac{\partial^2 u_i u_j}{\partial x_i \partial x_j}$ among many others.

The term *nonlocality* is used here in several related meanings which will become clear in the course of the discussion of the issues throughout this section and in the sequel. This includes anisotropy especially in small scales, statistical dependence of small and large scales, nonlocality versus decompositions, intermittency and structure(s), helicity, flows with additives, memory effects, flow history and predictability and effects of boundary including inflow conditions, closures and nonlocal relations between resolved and unresolved scales and constitutive relations and some other related issues. These are described, in Tsinober (2009, Sects. 1.3.5 and 6.6). Here we concentrate on the issues which have a touch to paradigmatic issues with some updating.

Nonlocality is a generic internal property of turbulent flows and exists independently of the presence of mean shear or other external factors, but has different manifestations for different external factors. For example, in the presence of a mean shear the small scales become anisotropic, whereas if the small scales are artificially excited the overall dissipation and mixing rate of the turbulent flow increase substantially. The direct interaction/coupling of large and small scales is in full conformity and is the consequence of the generic property of Navier–Stokes equations, which are integro-differential. It appears that the Kolmogorov 4/5 law can be interpreted as one of the manifestations of nonlocality in the above sense, see Sect. 8.1 below. Nonlocality is associated also with 'kinematics' due to the nonlocal relations between, e.g. velocity and its increments and between vorticity and strain. The non-local nature of the inertial nonlinearity is clearly seen looking at purely Lagrangian setting in which the inertial nonlinearity consists of the term with pressure only. We remind that from the formal point of view a process is called local if all the terms in the governing equations are differential. If the governing equations contain integral terms, then the process is nonlocal. The Navier–Stokes equations are *integro-differential* for the velocity field in both physical and Fourier space or any other. Therefore, generally, the Navier–Stokes equations describe nonlocal processes. Indeed, since $\nabla^2 p = \rho(\omega^2 - 2s_{ij}s_{ij}) = -\rho\frac{\partial^2 u_i u_j}{\partial x_i \partial x_j}$, the relation of pressure and velocity is nonlocal due to the nonlocality of the operator ∇^{-2}. This aspect of nonlocality is strongly associated with the essentially non-Lagrangian nature of pressure. Therefore it is not surprising that replacing in the Euler equations the pressure Hessian $\Pi_{ij} \equiv \frac{\partial^2 p}{\partial x_i \partial x_j}$, which is both nonlocal and non-Lagrangian, by a local quantity $\delta_{ij}\nabla^2 p = \rho/2\{\omega^2 - 2s_{ij}s_{ij}\}$ turns the problem into a local and integrable one and allows to integrate the equations for the invariants of the tensor of velocity derivatives $\partial u_i/\partial x_j$ in terms of a Lagrangian system of coordinates moving with a particle, see Cantwell (1992), Meneveau (2011) and references therein. One of the reasons for the disappearance of turbulence (and formation of singularity in finite time) in such models, called restricted Euler models, is that the eigenframe of s_{ij} in these models is *fixed* in space Novikov (1990b), whereas in a real turbulent flow it is oriented randomly in space and time. This means that nonlocality due to pressure is essential for (self-)sustaining turbulence: no pressure Hessian—no turbulence.

In view of special/paradigmatic status of the strain and enstrophy production it is of interest to mention that the pressure Hessian Π_{ij} enters the equations for the rate of change of strain and enstrophy production, i.e. $D(s_{ij}s_{jk}s_{kj})/Dt$ and $D(\omega_i\omega_j s_{ij})/Dt$ respectively, in the form $s_{ij}s_{jk}\Pi_{ij}$, $\omega_i\omega_j\Pi_{ij}$ which reflect the interaction between strain and vorticity and the pressure Hessian with a nonlocal contribution stemming from the deviatoric part of the pressure Hessian.

The fluid particle acceleration $\mathbf{a} \equiv D\mathbf{u}/Dt$—a kind of small scale quantity, which is dominated by the pressure gradient, ∇p and thereby the fluid particle acceleration is also related in a nonlocal manner to the velocity field. Hence the scaling properties of the acceleration variance do not obey Kolmogorov-like scaling. From the kinematic point $\mathbf{a} \equiv D\mathbf{u}/Dt = \partial\mathbf{u}\backslash\partial t + (\mathbf{u} \cdot \nabla)\mathbf{u} = -\frac{1}{\rho}\nabla p + \nu\nabla^2\mathbf{u} = \partial\mathbf{u}\backslash\partial t + \boldsymbol{\omega} \times \mathbf{u} + \nabla(u^2/2)$, i.e. it is a 'mixed' quantity due to presence of both velocity and velocity derivatives. Thus the impact of nonlocality on the behavior of acceleration.

There is a large set of fluid flows with "stronger" nonlocality. These are flows possessing additional mechanisms as the ability of supporting waves such as rotating, stably stratified and magnetohydrodynamic. Flows with special properties related to helicity such as with helical forcing are of this kind too.

Though turbulence is inherently nonlocal there are attempts to single out at least some aspects which are in some sense "local" by putting forward some hypotheses.

The main reason for such attempts becomes clear because nonlocality is among the main reasons of the absence of a sound theory of turbulence based on first principles. This state of matters is not unique for turbulence. Landau (1960) wrote: *It is well known that theoretical physics is at present almost helpless in dealing with the problem of strong interactions... and that it is necessary to consider "distributed", non-local, interactions... Unfortunately, the non-local nature of the interaction renders completely useless the technique of the present existing theory.*

It is probably one of the reasons why Kolmogorov (1985) wrote: *I soon understood that there was little hope of developing a pure, closed theory, and because of absence of such a theory the investigation must be based on hypotheses obtained on processing experimental data.*

Other reasons are more "practical". With nonlocality it is far from trivial if not impossible to use the experimental data—which are all limited in space and time—for "validation" of theoretical developments for, e.g. homogeneous flows, i.e. in 'infinite' domains. Also, locality is necessary for the 'physical foundation' of large-eddy simulation (LES) modeling of turbulence (Aluie 2012 and references therein).

Even just looking at the equations for the small/unresolved scales it is straightforward to realize that the small/unresolved scales depend on the large/resolved scales via nonlinear space and history-dependent functionals, i.e. essentially non-local both spatially and temporally. So it is unlikely—and there is accumulating evidence for this—that relations between them (such as "energy flux") would be approximately local in contradiction to K41a hypotheses and surprisingly numerous efforts to support their validity.

Before proceeding on the subject of nonlocal nature of turbulence several comments are given here on the issue of relevance and utility of locality in basic turbulence research. The first attempt of this kind was made by von Karman (1943) and von Karman and Howarth (1938) as an attempt to "solve" the famous Karman Howarth (KH) equation, containing both second and triple order correlations. This required a hypothesis to close the KH equation which is known as the so-called self-preservation (or self-similarity) hypothesis. The main basis for such a closure is the assumption that the turbulent motion at some point in time and space are defined by its immediate proximity. Hence locality. Though from the fundamental point of view there is little justification for the above closure, there is quite a bit of publications on the issue and related, see Sect. 16 in Monin and Yaglom (1975) and George (2012).

The next attempt of this kind was made by Kolmogorov (1941a) putting forward the statistical hypothesis of local homogeneity and isotropy: *...we think it rather likely that in an arbitrary turbulent flow with sufficiently large Reynolds number $Re = \frac{LU}{\nu}$ the hypothesis of local isotropy is realized with good approximation in sufficiently small regions... not lying close to the boundaries of the flow or its other special regions.* It is noteworthy that the hypothesis was formulated for *an arbitrary (i.e. not homogeneous isotropic, etc.) turbulent flow with sufficiently large Reynolds number $Re = \frac{LU}{\nu}$*. It was called by Batchelor (1953) as "universal equilibrium theory"; the term "equilibrium" was not given much justification and is causing a bit of confusion until now, see, e.g. references in George (2012) and also Jimenez (2012).

Locality is assumed in all treatments of "cascades" starting from Onsager (1945, 1949), for later references see Aluie (2012), Jimenez (2012) and references therein. This is done via employment of some decomposition and essential assumptions on existence of inertial range as defined by Kolmogorov (1941a) and locality of interactions in the cascade between components of similar magnitude most of which concern on how local is the energy transfer (spectral energy flux) with invariable claims on its locality in statistical sense in spite of the general expectation on nonlocality of interactions, which was convincingly demonstrated by Laval et al. (2001) on an example of a particular decomposition of the flow field. We mention also the so called "local equilibrium approximation" in wall bounded flows such that "statistically energy is dissipated close to where it is produced" at some locations along the normal to the wall (McKeon and Morrison 2007; Marusic et al. 2010; Smits et al. 2011; Jimenez 2012). As mentioned the motivation for locality at least in some sense is understandable, see above the quotation by Landau (1960), but the reality—which is the consequence of the nonlocal nature of turbulence—does not seem to fit mostly the predominantly wishful thinking on this issue. This is seen to some extent from what follows in the sequel. The bottom line is that generally locality cannot be assumed due to the inherently non-local nature of turbulence. For more on the issue of nonlocality as discussed above, see Tsinober (2009, Sect. 6.6, pp. 163–182). To put this differently, locality belongs to the same category as "small parameters", etc. No theory based on some locality, small parameters, etc., has little chance to succeed if any.

7.3.2 A Simple Example

We start from a simple example. Taking the position that velocity fluctuations represent the large scales and the velocity derivatives represent the small scales, one can state that, in homogeneous (not necessarily isotropic) the large and the small scales do not correlate. This can be expressed quantitatively by a correlation between velocity and vorticity. For example, in a homogeneous turbulent flow the Lamb vector

$\langle \boldsymbol{\omega} \times \mathbf{u} \rangle = 0$ and also $\langle (\mathbf{u} \cdot \nabla)\mathbf{u} \rangle = 0$. If the flow is statistically reflectionally symmetric, then $\langle \boldsymbol{\omega} \cdot \mathbf{u} \rangle = 0$ too. However, as mentioned, vanishing correlations do not necessarily mean absence of dynamically important relations. Indeed, the quantities $(\mathbf{u} \cdot \nabla)\mathbf{u} \equiv \boldsymbol{\omega} \times \mathbf{u} + \nabla(u^2/2)$ and $\boldsymbol{\omega} \times \mathbf{u}$, are the main 'guilty parties' responsible for all we call turbulence. Both contain the large scales (velocity) and small scales (velocity derivatives, strain, vorticity). So some kind of essential coupling between the two is more than unavoidable.

To illustrate the dynamical nature of this coupling we consider a unidirectional *in the mean* fully developed turbulent shear flow, such as the flow in a plane channel in which all statistical properties depend on the coordinate normal to the channel boundary, x_2, only. In such a flow, a simple precise kinematic relation is valid

$$d\langle u_1 u_2 \rangle / dx_2 \equiv \langle \boldsymbol{\omega} \times \mathbf{u} \rangle_1 = \langle \omega_2 u_3 - \omega_3 u_2 \rangle \neq 0,$$

which is just a consequence of the vector identity $(\mathbf{u} \cdot \nabla)\mathbf{u} \equiv \boldsymbol{\omega} \times \mathbf{u} + \nabla(\frac{u^2}{2})$ in which incompressibility and $d\langle \cdots \rangle / dx_{1,3} = 0$ where used, and $\langle \cdots \rangle$ means an average in some sense (e.g. time or/and over the planes $x_2 = const$, etc.). The *dynamic* aspect is that in turbulent channel flows $d\langle u_1 u_2 \rangle / dx_2 \neq 0$ is essentially different from zero at *any arbitrarily large* Reynolds number as far as the data allow to make such a claim, see Fig. 17 in Wei and Willmarth (1989). Therefore one can see from the above equation that at least some correlations between velocity and vorticity in such flows are essentially different from zero, see also Priyadarshana et al. (2007) and references therein. The important point is that without these correlations the mean flow would not "know" about the fluctuations at all whatever small are the corresponding correlation coefficients.

Let us look at the properties in the proximity of the midplane, $x_2 \approx 0$, of the turbulent channel flow. In this region $dU/dx_2 \approx 0$ and $\langle u_1 u_2 \rangle \approx 0$, but—contrary to common assumptions *the flow is neither homogeneous nor isotropic also at level of velocity derivatives*, since the gradient $d\langle u_1 u_2 \rangle / dx_2$ is essentially $\neq 0$ and is *finite independently of the Reynolds number*. This is also a clear indication of nonlocality, since in the bulk of the flow, i.e. *far from the boundaries*, $dU/dx_2 \sim 0$, and also a clear counterexample to the hypothesis of the local isotropy: even in the proximity of the centerline of the channel this hypothesis does not hold for whatever large Reynolds numbers. The absence of the $k^{-5/3}$ spectrum at the highest available Reynolds numbers, $Re = 230000$, based on the mean velocity and half channel width (Compte-Bellot 1965) is consistent with the above. It is noteworthy that presence of two walls is essential.

7.3.3 Direct and Bidirectional Coupling Between Large and Small Scales

Here we concentrate on nonlocality as manifested in direct and bidirectional interaction/coupling between large and small scales. There exist massive evidence that this is really the case as there are many indications that this interaction is bidirectional. We should first mention the well known effective use of fine honeycombs and screens in reducing large scale turbulence in various experimental facilities (Laws and Livesey 1978; Tan-Attichat et al. 1989). The experimentally observed phenomenon of strong drag reduction in turbulent flows of dilute polymer solutions and other drag reducing additives is another example of such a 'reacting back' effect of small scales on the large scales. Third, one can substantially increase the dissipation and the rate of mixing in a turbulent flow by *directly* exciting the small scales experimentally, e.g. in a jet and in DNS in a periodic box (Suzuki and Nagano 1999; Vukasinovich et al. 2010). Similar effects are observed with small-scale acoustic excitation. Fourth, mostly recent developments in wall bounded shear flows revealed the importance of interactions of structure(s) in the flow, such as inner-outer interactions including the large-small scales, generally, and with the nearwall region, in particular (Klewicki 2010; Marusic et al. 2010; Smits et al. 2011; Jimenez 2012).

As one of the premises we would like to remind that vorticity and strain are not just velocity derivatives. They are special for several reasons. As discussed above they reflect the rotational and dissipative nature of turbulence. In the context of nonlocality the property to be stressed here is that the whole flow field is determined entirely by the field of strain and/or vorticity with appropriate boundary conditions: $\nabla^2 u_i = 2\partial s_{ij}/\partial x_j$; $\nabla^2 \mathbf{u} = -\operatorname{curl}\boldsymbol{\omega}$; i.e. the velocity field is a linear functional of strain $u_i = \mathfrak{G}\{s_{ij}\}$ and/or vorticity $\mathbf{u} = \mathfrak{F}\{\boldsymbol{\omega}\}$, i.e. the field of velocity $\mathbf{u}(\mathbf{x}, t)$ in each space point $\mathbf{x}$ is defined by the whole field of strain and/or vorticity. Alteration of the field of velocity derivatives reflects on the velocity field, vorticity and strain are not passive—they react back and not only for the above (kinematic) reason. This in turn means that the large scales as represented by the velocity field and the small scales as represented by vorticity and strain, should be strongly coupled (which is not the same as being correlated), as indeed is the case.

Along with the kinematic relation above $u_i = \mathfrak{G}\{s_{ij}\}$ (and $\mathbf{u} = \mathfrak{F}\{\boldsymbol{\omega}\}$) the production of strain $-s_{ij}s_{jk}s_{kj}$ (and enstrophy $\omega_i\omega_j s_{ij}$) 'reacts back' in creating the corresponding velocity field, i.e. the small scales are not just 'swept' by the large ones. Therefore, it is incorrect to treat the small scales as a kind of passive object (e.g. passive sink of energy) just 'slaved' to the large scales (e.g. the velocity field), which are not decoupled from the small ones just like the vice versa. Moreover, at the level of velocity derivatives the inertial, $-s_{ij}s_{jk}s_{kj}$, $\omega_i\omega_j s_{ij}$ and viscous terms, $\nu\omega_i\nabla^2\omega_i$, $\nu s_{ij}\nabla^2 s_{ij}$, do not act as if they were additive and independent—their interaction is crucial and among other things is manifested in the (approximate) Tennekes and Lumley (1972) balance for the enstrophy production and a similar relation for strain production, see Sect. 6.3 in Tsinober (2009).

For a statistically stationary and homogeneous flow with large scale forcing

$$0 \approx \langle \omega_i\omega_j s_{ij}\rangle - \nu\big\langle(\partial\omega_i/\partial x_k)^2\big\rangle; \qquad 0 \approx -(2/3)\langle s_{ij}s_{jk}s_{ki}\rangle - \nu\big\langle(\partial s_{ij}/\partial x_k)^2\big\rangle.$$

Using simple estimates for a turbulent shear flow at high Reynolds number Tennekes and Lumley (1972) arrived to a similar approximate balance as the first relation above called Tennekes and Lumley balance (TL). An important aspect is that this balance appears to be valid in different meanings (not only in the mean) as follows. The first feature that the TL balance holds at Re_λ as low as ≈ 60. Second, it holds pointwise in time, i.e. the integrals over the flow domain of the enstrophy production and of its viscous destruction are approximately balanced at any time moment, $\int \omega_i \omega_j s_{ij} dV \approx -\nu \int \omega_i \nabla \omega_i dV$. The features concerning the TL balance appear to be true for temporally modulated turbulent flows and for flows with hyper-viscosity of different orders, $h = 2, 4, 8$, $h = 1$ corresponds to Newtonian fluid. The approximate balance as described above between the 'inertial' and diffusive processes shows that they are very far from being additive and point to strong mutual interaction. In particular, this is a strong indication that the nature of dissipation is important in the enstrophy/strain production and the properties of the vorticity/strain in the whole flow field. This in turn means that it should be important in the properties of the velocity too as the latter is fully determined by the field of vorticity/strain. We return to these issues in section Chap. 7 below, also for more on this see Tsinober (2009, Chap. 6, pp. 135–141).

It is noteworthy, that one of the claims is that the TL balance is the reason is the for the predominant enstrophy (and strain) production. This argument is misleading and puts the consequences before the reasons, since it is known that, for Euler equations, the enstrophy production increases with time very rapidly, apparently without limit, see references in Tsinober (1998a, 1998b, 2000). Another rather common view that the prevalence of vortex stretching is due to the predominance of stretching of material lines which is due to an erroneous, for more on the differences between these processes see Tsinober (2009, Chap. 9, pp. 307–310).

In other words, the presence of viscosity changes qualitatively the nature of the enstrophy/strain production and the properties of the vorticity/strain field. This in turn means that the nature of dissipation/viscosity is important in the properties of the velocity (including structure functions, inertial range, etc.) since—as mentioned—the latter is fully determined by the field of vorticity/strain. In other words, the nonlocal impact of the small (including dissipative) scales on the whole flow field is realized through the chain $-s_{ij}s_{jk}s_{kj}$, $\omega_i\omega_j s_{ij}$ to ω^2, s^2 (and ω_i, s_{ij}) and the velocity field as described above, including what can be termed the back reaction of dissipation on the whole flow field, see Chap. 7 below.

A noteworthy aspect is the nonlocality of (purely kinematic) relations between the fields of vorticity and strain and similar relations such as between $-s_{ij}s_{jk}s_{kj}$ and $\omega_i\omega_j s_{ij}$, see Fig. 7.2.

The above and other aspects of nonlocality contradict the idea of cascade in physical space, which is local by definition, e.g. see Frisch (1995, p. 104). For example, the commonly assumed statistical independence between large and small scales (i.e. sweeping decorrelation hypothesis) both of (i) structure functions $S_p \equiv \langle (\Delta u)^p \rangle$; $\Delta u \equiv \mathbf{u}(\mathbf{x} + \mathbf{r}) - \mathbf{u}(\mathbf{x}\mathbf{r}/r)$ on the (nature) of dissipation, i.e. strain, in the 'inertial range' and (ii) the small scales on the large ones stand in contradiction with the relation $u = \mathfrak{G}\{s_{ij}(x,t)\}$ together with the process of self-production of strain in turbulent flows. This leads to ill-posedness of both the concepts of cascade and inertial range with the former being the reason for the so called anomalous scaling, see Chap. 8.

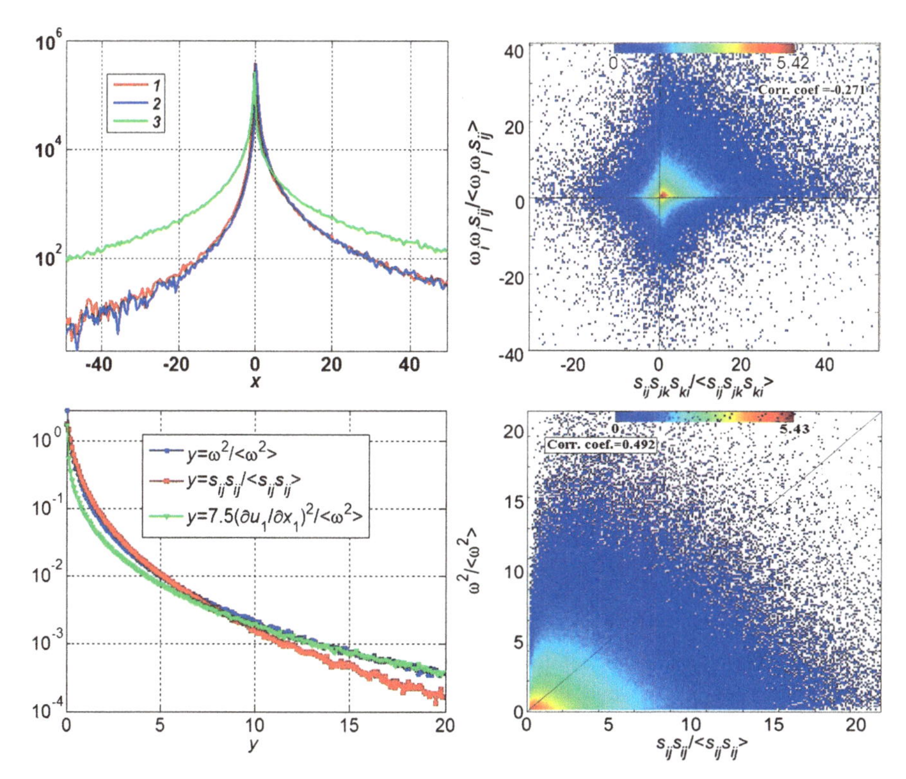

Fig. 7.2 *Top*: *Left*—PDFs of $3/4\omega_i\omega_j s_{ij}$, $-s_{ij}s_{jk}s_{ki}$, and $-17.5(\partial u_1/\partial x_1)^3$ normalized on their means, $Re_\lambda = 10^4$, *Right*—Joint PDF and scatter plot of $3/4\omega_i\omega_j s_{ij}$ versus $-s_{ij}s_{jk}s_{ki}$, normalized on their means. Note that though the univariate PDFs of $3/4\omega_i\omega_j s_{ij}$ and $-s_{ij}s_{jk}s_{ki}$ are practically the same, their Joint PDFs show that the strongest activity in strain production corresponds to weakest that of enstrophy and vice versa, which is not the case for the strain and enstrophy themselves as seen from the figure on bottom right. *Bottom*: *Left*—PDFs of ω^2 and s^2, *Right*—their Joint PDF. Field experiment at, $Re_\lambda = 10^4$ (Gulitskii et al. 2007a, 2007b, 2007c)

Direct and bidirectional coupling between large and small scales is observed in diverse manifestations. These are described in Sect. 6.6. in Tsinober (2009). One of the issues concerns the statistical dependence of small and large scales. The impact of large scales on the small ones is known for a while. It was addressed by Kolmogorov (1962) following the famous remark by Landau (1944)[3] and some earlier experimental data (Monin and Yaglom 1971; Obukhov 1962). Kolmogorov formulated modified K41 hypotheses assuming among other things the log-normal

[3]Kolmogorov did these modifications following the Landau objection to universality in the first Russian edition of Fluid Mechanics by Landau and Lifshits (1944) about the role of large-scale fluctuations of energy dissipation rate, i.e., non-universality of both the scaling exponents and the prefactors: *important part will be played by the manner of variation of ϵ over times of the order of the periods of large eddies (of size ℓ)*, see Landau and Lifshits (1987, p. 140).

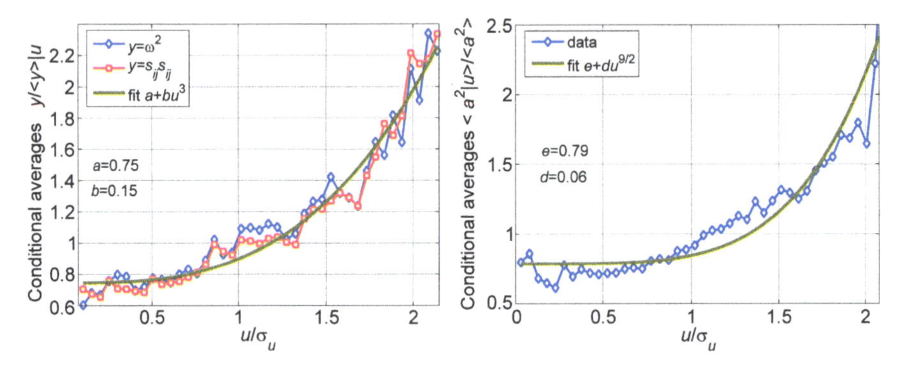

Fig. 7.3 *Left*—conditional averages of enstrophy ω^2 and total strain $s_{ij}s_{ij}$ conditioned on magnitude of velocity fluctuations vector, u. The fit is in the spirit of the Kolmogorov refined similarity hypothesis, though it is a fit in the first place. This fit cannot be expected to be universal quantitatively and should at least have different coefficients a and b for flows with different large-scale properties in the spirit of the Landau remark. *Right*—conditional averages of squared acceleration magnitude a^2 on magnitude of velocity fluctuations vector, u (Gulitskii et al. 2007a, 2007b)

distribution coarse-grained dissipation,[4] i.e. averaged over a sphere of radius r, with r in the conventionally defined inertial range (CDIR), i.e. $L > r > \eta$ with some 'external scale' L and the 'internal scale' $\eta = (\nu^3/\langle\epsilon\rangle)^{1/4}$. The main result is that the statistics of structure functions in the CDIR depends on the large scale structure of the flow both in scaling exponents and prefactor: $S_p \equiv \langle(\Delta u)^p\rangle = C_p(\mathbf{x},t)(L/r)^{\frac{1}{2}kp(p-3)}(r\langle\epsilon\rangle)^{p/3}$; for $p=2$ giving $S_2 = C_p(\mathbf{x},t)(L/r)^{-k}(r\langle\epsilon\rangle)^{2/3}$. The factor $C(\mathbf{x},t)$ "depends on the macrostructure of the flow", k is the so called "intermittency" exponent. This model captures qualitatively the correction to the structure function scaling, though it is known that turbulence is not lognormal.

Flows with inhomogeneous large scales, large scale or mean strain, shear etc. can exert stronger impact on small scales—the issue goes back to Corrsin (1958), see references, in Saddoughi (1997), Hill (2006) and Tsinober (2009). But there is strong statistical dependence of small scales on the large ones in weakly-inhomogeneous/anisotropic flows either, e.g. Fig. 7.3.

A less trivial is another aspect which until recently was neglected as unimportant: the impact of small scales, especially in the dissipative range with $r < \eta$, on the conventionally defined inertial range (CDIR), which appears to be contaminated also by strong dissipative events from the conventionally defined dissipative range (CDDR). It appears that quite the opposite is true as found in recent experiments at large Reynolds numbers up to $Re_\lambda \approx 10^4$ with access to the tensor of velocity

[4]This assumption is due to Obukhov (1962) because as he wrote it *not very restrictive as an approximate hypothesis since the distribution of any essentially positive characteristic can be represented by a logarithmically Gaussian distribution with correct values of the first two moments.*

This is correct for empirical purposes, but when it goes about the right results for the right reasons it is not sufficient.

derivatives $\partial u_i / \partial x_j$ and in particular local strain (i.e. dissipation) and vorticity, see Borisenkov et al. (2011), Kholmyansky and Tsinober (2009), Tsinober (2009) and references therein. The main underlying reason is again nonlocality. We address this issue in the next section for several reasons the main being that it is of special importance at large Reynolds numbers.

Chapter 8
Large Reynolds Number Behavior, Symmetries, Universality

Abstract The large Reynolds number behavior is of special importance from several points of fundamental nature which include such issues as restoring (or not) the symmetries of Navier–Stokes equations, and in some sense even scale invariance of Euler equations via the "multifractal formalism" (Frish, Turbulence: the legacy of A.N. Kolmogorov, 1995, pp. 18, 144), and universality, the role of viscosity/dissipation and the concept of inertial range, the role of the nature of forcing/excitation, inflow, initial and boundary conditions. In view of the arguments and experimental results on nonlocality and the direct and bidirectional coupling between large and small scales in the previous section a natural question arises what is the impact of nonlocality on all the above and whether there are enough reasons and evidence for a discussion and reexamination of the above issues, generally, and in relation to the nonlocal properties of turbulent flows among others, especially. Navier–Stokes equations at sufficiently large Reynolds number have the property of intrinsic mechanisms of becoming complex without any external aid including strain and vorticity amplification. There is no guarantee that the outcome is the same from, e.g. natural "self-randomization" and with random forcing, on one hand, and different kinds of forcing, boundary and initial conditions, on the other hand. Moreover there is serious evidence that the outcome may be and indeed is different.

We have chosen to start with the issue of the role of viscosity and/or dissipation (including their nature) and the concept of inertial range as most convincing to demonstrate that the concern is justified. Indeed, there are every reasons not only to expect, but also hard experimental evidence that the small scales from the dissipative range, at least those which are strong, have an important impact on the larger scales including the conventionally defined inertial range (CDIR). Among the main points below is that due to nonlocality at least some key nonlinear terms are not purely inertial in the CDIR in contrast to common beliefs, the 4/5 Kolmogorov law at large *Re* being an outstanding example of not purely inertial relation and a victim of interpretational abuse.

We have chosen to start with the issue of the role of viscosity and/or dissipation (including their nature) and the concept of inertial range as most convincing to demonstrate that the concern is justified. Indeed, there are every reasons to expect that the small scales from the dissipative range, at least those which are strong, should have

A. Tsinober, *The Essence of Turbulence as a Physical Phenomenon*,
DOI 10.1007/978-94-007-7180-2_8,

important impact on the larger scales including the conventionally defined inertial range (CDIR). Among of the main points below is that due to nonlocality at least some key nonlinear terms are not purely inertial in the CDIR.

8.1 Inertial Range, the Roles of Viscosity/Dissipation and Related Issues

We start with reminding the common view that turbulence is an essentially inertial phenomenon such that how the energy is dissipated in the small scales does not influence the large scales as long as the amount is correct, i.e. the small scales are just a passive sink of energy and the nature of dissipation does not matter for large scales and in particular in the CDIR,[1] see references and citations on this kind of statements at pp. 103 and 335 in Tsinober (2009). Some more are given in the Appendix essential quotations.

One of the popular arguments for the existence of the conventionally defined inertial range (CDIR) at large Reynolds numbers is the Kolmogorov 4/5 law which is the consequence of the NSE and under some assumptions (including isotropy) takes the following form, for a precise form, see Eq. (34.20), p. 139 in Landau and Lifshits (1987),

$$S_3 = -(4/5)\langle\epsilon\rangle r + 6\nu dS_2/dr$$

and in which the last term is negligible at large Reynolds numbers. Moreover, there is an experimental confirmation that at Reynolds numbers up to $Re_\lambda \approx 10^4$ the term $6\nu dS_2/dr$ is negligible so that the relation

$$S_3 = -(4/5)\langle\epsilon\rangle r,$$

does hold in a broad range of scales exceeding three decades at $Re \approx 10^4$ (Kholmyansky and Tsinober 2008).

Thus one gets an impression that the 4/5 is a purely inertial relation at large Re, and that the third-order moment is universal, i.e. it does not depend on the details of the turbulence production, but is determined solely by the mean energy dissipation rate only. However, this would be true if the $S_3 \equiv \langle(\Delta u)^3\rangle$, or more precisely just velocity increments $\Delta u \equiv [\mathbf{u}(\mathbf{x}+\mathbf{r}) - \mathbf{u}(\mathbf{x})]\mathbf{r}/r$, would *not* contain non-negligible contributions from dissipative events, i.e. *in reality what is called IR* is ***assumed to be purely inertial and having no contribution from viscous events***. Indeed, computing Δu one encounters also large and even very large instantaneous dissipation at the ends $(\mathbf{x}, \mathbf{x}+\mathbf{r})$.[2] In other words, the second Kolmogorov (1941a) hypothesis involves a strong assumption that the dissipative events such that at least at one of

---

[1]This latter is due to Kolmogorov (1941a). We emphasize that the correction he proposed in 1962 concerns the impact of the large scales on the scaling of structure functions in CDIR.

[2]The same happens, e.g. when computing the increments of Lagrangian velocity along the fluid particle trajectory. The consequences are expected to be the same at least qualitatively.

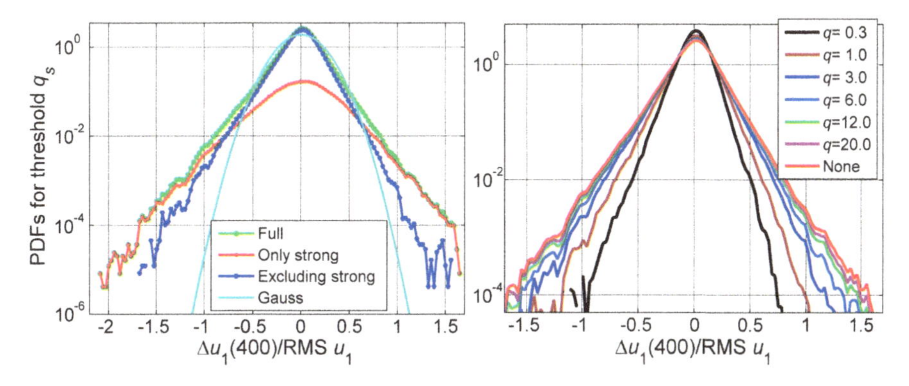

Fig. 8.1 Histograms of the increments of the longitudinal velocity component for the full data and the same data in which the strong dissipative events were removed: (**a**) $r/\eta = 400$ and threshold $q = 3$, (**b**) $r/\eta = 400$ with different thresholds; field experiment, $Re_\lambda \approx 10^4$;—the lower edge of the inertial range is about $r/\eta = 40$. An event $\Delta u \equiv [u(x+r) - u(x)]r/r$ is qualified as a strong dissipative if at least at one of its ends $(x, x+r)$ the instantaneous dissipation $\epsilon > q\langle\epsilon\rangle$ with $q > 1$. Note that the PDFs with removed strong dissipative events (*dark blue* ones) are not close to the Gaussian curve. An important feature is that dissipative events literally live within the CDIR in turbulence at high Reynolds numbers, i.e. nonlocality is a broader issue and does not necessarily involve scale separation such as in long range interactions (Kholmyansky and Tsinober 2009)

their ends $(\mathbf{x}, \mathbf{x}+\mathbf{r})$ the instantaneous dissipation $\epsilon > q\langle\epsilon\rangle$ with $q > 1$ do not matter for the statistics of velocity increments. To (dis)prove this one needs access to instantaneous dissipation at large Reynolds numbers. Indeed, looking at Fig. 8.1 it is seen that there exists a substantial number of dissipative (!) events (DE) living in the conventionally defined dissipative range (CDDR) with contributing essentially to the PDF of velocity increments in the conventionally defined inertial range (CDIR) at high Reynolds numbers, $Re_\lambda \approx 10^4$.

The key feature is that this contribution is largest to the tails of the PDF of velocity increments. Thus the CDIR is an ill-defined concept. In particular, this means that the neglected viscous term in the Kolmogorov 4/5 law does not contain all the viscous contributions. Those present in the structure function S_3 itself remain and keep the 4/5 law precise: without the dissipative events just mentioned the 4/5 law does not hold! In this sense the 4/5 law is not a pure inertial law even at $Re_\lambda \approx 10^4$. Indeed, strong dissipative events do contribute to the 4/5 law, see Fig. 8.2, and removing them leads to an increase of the scaling exponent above unity, see insert in Fig. 8.2. It is noteworthy that the contribution of the dissipative events in the 4/5 law at large Reynolds numbers is not small in spite of considerable cancellation between the negative and positive events.

Thus—contrary to common view—the 4/5 law is not a pure inertial relation at large *Re*. The important implication of more general nature is that nonlinear interactions (NI) are not synonymous to purely inertial ones. They (NI) consist of purely inertial ones with an essential contribution from the viscous and cross-interactions. In case of the third order functions and energy fluxes these interactions are constrained by the 4/5 law. The consequence is that the purely inertial and dissipative events

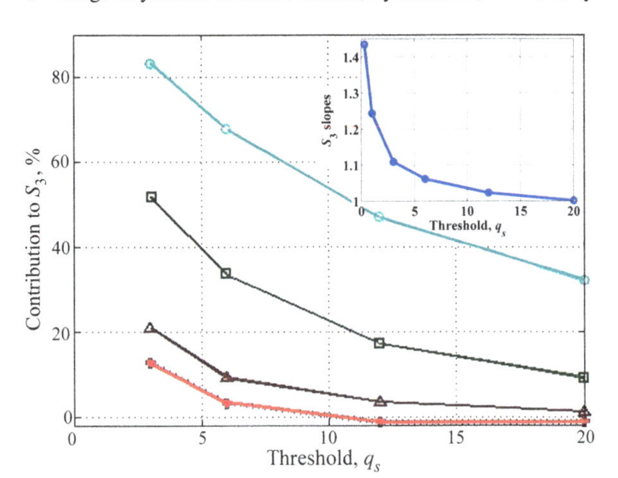

Fig. 8.2 Contributions of the strong dissipative events in the CDIR to the third-order structure function as a function of the threshold q for various separations r; *blue circles*—$r/\eta = 4$, *green squares*—$r/\eta = 40$, *brown triangles*—$r/\eta = 400$, *red crosses*—$r/\eta = 4000$; in the *insert*: scaling exponents of the third-order structure function as a function of the threshold (Kholmyansky and Tsinober 2009)

are adjusting to keep some quantities such as the total dissipation or the energy flux approximately constant at large Reynolds numbers. The constraint responsible for this adjustment is just the 4/5 law or more generally the NSE.[3]

In other words, the independence of some parameter of viscosity at large Reynolds numbers does not mean that viscosity is unimportant. It means only that rather than being unimportant the (cumulative) effect of viscosity is Reynolds number independent. Thus, speaking about asymptotic (large *Re*) behavior a realistic option is that some (not all) quantities $\Longrightarrow$ *const* as they do in observations, but this does not mean that they have to become independent on viscosity, i.e. generally the limit $\nu \Longrightarrow 0$ is not viscosity independent. We return to this issue below, for more on this issue see Sect. 10.2 in Tsinober (2009).

As concerns structure functions of higher order $p > 3$ there is no simple constraint as the 4/5 law for S_3. The consequence is that the mentioned above dissipative events are responsible for what is called anomalous scaling for S_p, $p > 3$. This is clearly seen from Fig. 8.3. Thus the anomalous scaling is not an attribute of the conventionally-defined inertial range (CDIR), and the latter is not a well-defined concept, just like "cascade" and the conventionally defined dissipative range, CDDR.

A noteworthy conjecture concerns the Lagrangian setting in which the sweeping effects are mostly (but not totally! see Chap. 4) removed so that the contribution of the strong dissipative events (SDE) in the CDIR is even larger than in the Eulerian setting. Thus one is tempted to bet that this is the reason for the absence of the expected scaling *a la* Kolmogorov 41 (it is due to Landau and Lifshits 1944) even at the level of the second order Lagrangian structure function. In other words removal

---

[3]A noteworthy is the pure kinematic relation involving the third order structure function (Hosokawa 2007; Kholmyansky and Tsinober 2008; Germano 2012), $-\langle(\Delta u)^3\rangle = 3\langle u_{sum}^2 \Delta u\rangle$, with $2u_{sum} = u(x+r) + u(x)$, which together with the 4/5 law results in a relation equivalent to the 4/5 law $\langle u_{sum}^2 \Delta u\rangle = \langle\epsilon\rangle r/30$. Thus the 4/5 law provides a clear indication of absence of statistical independence between the quantities residing at large and small scales.

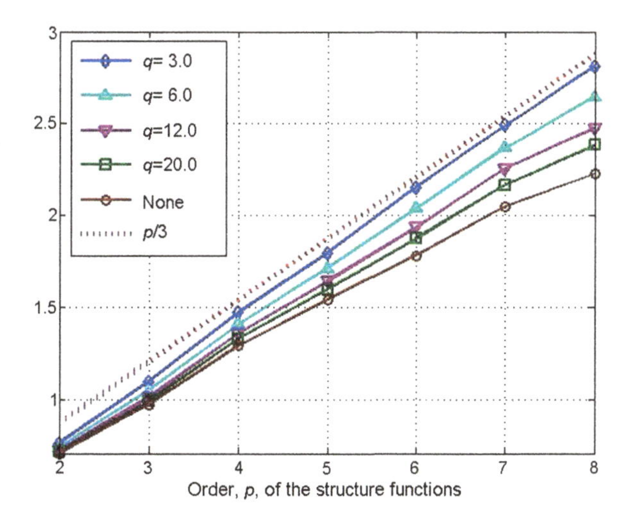

Fig. 8.3 Scaling exponents of structure functions at $Re_\lambda \approx 10^4$ for the longitudinal velocity component corresponding to the full data and the same data in which the strong dissipative events with various thresholds q were removed. With $q = 3$ the higher order structure functions ($p > 3$) exhibit Kolmogorov scaling $p/3$

of the SDE from appropriate Lagrangian data set as above in the Eulerian setting would bring the expected scaling, which is not observed so far (Falkovich et al. 2012).

It is naturally to expect a similar phenomenon of non-negligible contribution of dissipative events to quantities in the CDIR on a more general level than just structure functions. For example, not all quantities entering equations such as RANS or LES for resolved scales in the "inertial range" are indeed purely inertial. Indeed, it appears strong dissipative events make a nonnegligible contribution to the SGS stresses and SGS energy flux $\Pi(x; r) = -\tau_{ik}[s_{ik}]$ where $\tau_{ik} = [u_i u_k] - [u_i][u_k]$, are the SGS stresses and the filtered quantities are denoted as $[\ldots]$, see Fig. 8.4. Indeed, this is observed on a qualitative level by necessity of using time series and one dimensional standard Gaussian filter of width r. There is an essential contribution of the dissipative events to the SGS energy flux in the conventionally-defined inertial range with a considerable dependence on the threshold q, which is increased with the decrease of q. The main point is that the Reynolds stresses, subgrid stresses, etc. are not inertial range quantities, because the contribution of viscous effects is not limited by the viscous terms corresponding, e.g., to the Laplacian in the filtered equations as in the case of 4/5 law.

Another example of the contribution of DE within the CDIR is observed in the behavior of other key quantities such as the enstrophy and strain production $\omega_i \omega_j s_{ij}$, $s_{ij} s_{jk} s_{ki}$. Though the origin of both quantities and corresponding processes is purely inertial, it appears that both quantities contain a substantial contribution from the dissipative events. This is clearly seen from Fig. 8.5 showing the contribution of DE and examples of the PDFs of $\omega_i \omega_j s_{ij}$ and $s_{ij} s_{jk} s_{ki}$ for the whole data and the sets with removed dissipative events.

This is a direct demonstration of the nonlocal impact of the dissipative scales on the whole flow field through the chain $-s_{ij} s_{jk} s_{kj}$, $\omega_i \omega_j s_{ij}$ to ω^2, s^2 and the velocity field as described above. This includes what is termed the back reaction of dissipation on the whole flow field.

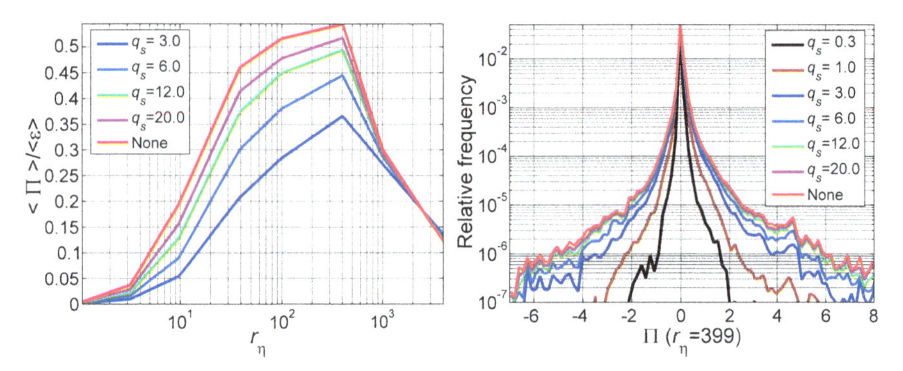

Fig. 8.4 An example showing the contribution of strong dissipative events to the subgrid scale (SGS) energy flux $\Pi(x;r) = -\tau_{ik}[s_{ik}]$ at $Re_\lambda = 10^4$ for the longitudinal velocity component corresponding to the full data and the same data in which the strong dissipative events (when at least at one point x or $x+r$ the instantaneous dissipation $\epsilon > q\langle\epsilon\rangle$) with various thresholds q were removed

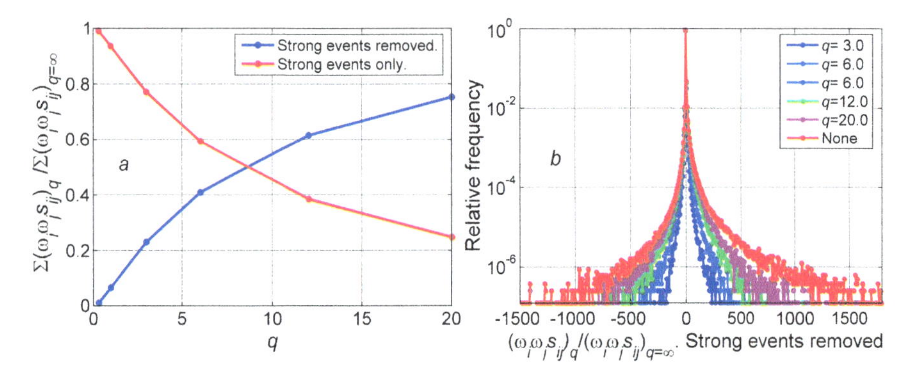

Fig. 8.5 (**a**) Contributions of strong dissipative events (when at point x the dissipation $\epsilon > q\langle\epsilon\rangle$) to the enstrophy production as a function of the threshold q. (**b**) Histograms of the enstrophy production for the same data with removed strong dissipative events for various values of the threshold q. Similar results are observed for strain production, see Fig. 4 in Borisenkov et al. (2011)

This impact of viscosity is seen also from the equations for strain and vorticity with the production terms $-s_{ij}s_{jk}s_{kj}$, $\omega_i\omega_j s_{ij}$ and is consistent with the Tennekes and Lumley balance, see Chap. 7 above. As mentioned an important aspect is that this balance appears to be valid in different meanings (not only in the mean) as follows. The first feature that the TL balance holds at Re_λ as low as ≈ 60. Second, it holds pointwise in time, i.e. the integrals over the flow domain of the enstrophy production and of its viscous destruction are approximately balanced at any time moment, $\int \omega_i\omega_j s_{ij} dV \approx -\nu \int \omega_i \nabla\omega_i dV$. The features concerning the TL balance appear to be true for temporally modulated turbulent flows and for flows with hyperviscosity of different orders, $h = 2, 4, 8$ with $h = 1$ corresponding to Newtonian fluid.

Before proceeding further a short summary of the above is as follows:

(A) The important aspects of the evidence presented just above is that the sub-Kolmogorov scales (dissipative events—DE), defined as conventionally dissipative range (CDDR), are directly and bidirectionally coupled with the conventionally-defined inertial range (CDIR). These DE literally live within the CDIR in turbulence at high Reynolds numbers, e.g. they make the largest contribution to the tails of the PDF of velocity increments in the CDIR. Hence several non-trivial consequences including those of paradigmatic nature: (i) both the CDIR and CDDR (and thereby cascade) are ill defined concepts, e.g. it is the presence of the dissipative events which is responsible for what is called anomalous scaling in the CDIR, i.e. its 'anomalous scaling' is not an attribute of the inertial range simply as such is not in existence (ii) the 4/5 law is a not pure inertial relation at large *Re*. Similarly filtered equations for the inertial range quantities are not purely inertial range relations either. In both there is an essential contribution of dissipative events from the conventionally dissipative range. The same is true of such processes as enstrophy and strain production and other key nonlinear processes, these are to not purely inertial processes with an essential contribution from the of dissipative events from the conventionally dissipative range.[4]

(B) In other words the recent experimental evidence with direct access to the field of velocity derivatives at large Reynolds numbers does not support the hypothesis on existence of an inertial range with some scaling symmetry at large Reynolds numbers with viscosity/dissipation being far more important than just a passive sink of energy. A similar state of matters is with the issue of universality and related. Many beliefs in various aspects of universality (quite understandable, especially needed in theoretical treatments) appear to be not real—the evidence points to the contrary: the flow, generally, depends on the nature of forcing, inflow, initial and boundary conditions and a variety of issues in flow control(liability) both in engineering and mathematical contexts. After all NSE, BCs and ICs define the flow field in its entirety.

The above calls for a discussion and reexamination of the issues mentioned above such as asymptotics at large Reynolds numbers, symmetries and universality especially in relation to the nonlocal properties of turbulent flows among others.

8.2 Reynolds Number Dependence and Behavior of Turbulent Flows at Large Reynolds Numbers

Observations show that there are two kinds of properties of turbulent flows. Some properties of turbulent flows become Reynolds number independent as the Reynolds

[4]Thus the question by Kraichnan (1974): *How a theoretical attack on the inertial-range problem should proceed is far from clear* seems to be irrelevant as there is no such an object in existence.

number becomes large enough. Other remain Reynolds number dependent and the existing modest evidence indicates that they may never saturate as $Re \to \infty$.

The first kind of properties is represented by the drag of bluff bodies (a circular disc is one of the cleanest examples—it's drag coefficient is independent of Reynolds number beyond $Re \sim 10^3$); another example is exhibited by the independence of dissipation of Reynolds number in a variety of flows form basic configurations to diverse applications in engineering, geophysical flows and others all of which are neither homogeneous nor isotropic. Things like 2/3, 4/5 and 4/15 laws, $k^{-5/3}$ spectrum and some others belong to the same category.

However, it should be stressed that the meaning of Reynolds independence (and being insensitive to the nature of dissipation) requires caution, since, in fact, there is a "hidden" dependence on viscous effects as with the 4/5 law even at $Re_\lambda \sim 4 \times 10^4$ as demonstrated above. In other words, the independence of some parameter of viscosity at large Reynolds numbers does not mean that viscosity is unimportant. It means only that the (cumulative) effect of viscosity is Reynolds number independent. Moreover, though the dissipation is known empirically to tend presumably to saturate to a nonzero limit as $\nu \to 0$ (or at least essentially non-vanishing) this however, does not mean that there exist a limit as $\nu \to 0$ in the sense (or any other sense) that all other flow characteristics do saturate as well.

As a simple illustration let us look at the consequence of dissipation being constant for the field of velocity derivatives. Since at large $Re \langle \epsilon \rangle = 2\nu \langle s^2 \rangle \approx \nu \langle \omega^2 \rangle$ this means that at large Reynolds numbers both $\langle s^2 \rangle$, $\langle \omega^2 \rangle \approx \nu^{-1}$, i.e. the field of velocity derivatives is not only Reynolds number dependent, but also becomes singular in the limit $\nu \to 0$ ($Re \to \infty$). Due to the intermittent nature of the field of velocity derivatives one can expect that the maximal values of s^2, ω^2 (or $\max(|\partial u_i / \partial x_k|)$) increase even faster with the Reynolds number. This possible unboundedness of the field of velocity derivatives as $Re \to \infty$ has an implication that the Newtonian approximation can break down as $\nu \to 0$, since the linear stress/strain relation is only the first term in the gradient expansion. So far, however, there seems to be no evidence that the Navier–Stokes equations are inadequate for describing turbulent flows, i.e. breakdown of the of the continuum approximation, which is also an indication of absence of breakdown of the NSE due to possible formation of singularities in finite time; but still it is safe to keep in mind that any equations are not Nature. In this context it is instructive to remind a quotation from Goldstein (1972) concerning the success of NSE equations for the laminar flows of viscous fluids, but even in this case, *it is, in fact, surprising that the assumption of linearity in the relation between* τ_{ij} *and* s_{ij} *as usually employed in continuum theory,... works as well, and over as large a range, as it does. Unless we are prepared simply to accept this gratefully, without further curiosity, it seems clear that a deeper explanation must be sought.* Among the possible reasons that the possible violation is not so easy to detect is that, even if it happens, it will occur at rather large Reynolds numbers and small regions due to the strong intermittency of the field of velocity derivatives.

Let us turn to the Re-dependence at large Reynold numbers. There exists considerable evidence on Reynolds number dependence of various properties in different turbulent flows; for a partial list of recent references, see Tsinober (2009). The first

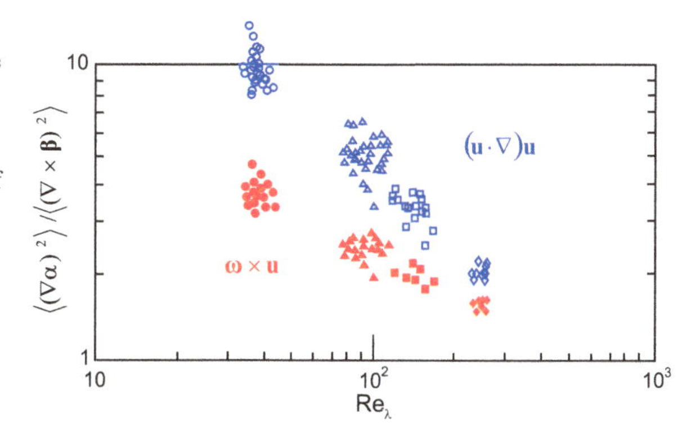

Fig. 8.6 Reynolds-number dependence of the ratio of the variances of the irrotational and solenoidal parts of the nonlinear term $(\mathbf{u}\cdot\nabla)\mathbf{u}$ and $\boldsymbol{\omega}\times\mathbf{u}$ in a DNS simulation of quasi-isotropic turbulence (Tsinober et al. 2001)

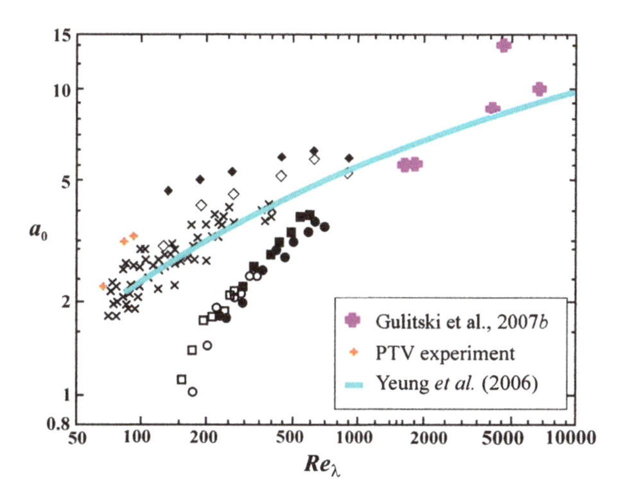

Fig. 8.7 Normalized acceleration variance, $a_0 = (1/3)a_k a_k^{3/2}\nu^{-1/2}$ versus Re_λ, from Gylfason et al. (2004) with added experimental data from field experiment (Gulitskii et al. 2007b) and from the PTV experiments (Lüthi et al. 2005)

example is the flatness factor of the streamwise velocity derivative $\partial u_1/\partial x_1$ is increasing from 3–4 at $Re_\lambda \sim 10$ to about 40 at $Re_\lambda \sim 4\times 10^4$, without showing any trend for saturation, see Fig. 6 in Sreenivasan and Antonia (1997); (Gulitskii et al. 2007a and references therein). There is no understanding of the reasons for such a strong Reynolds number dependence at large values of the Reynolds number. Another example is about the Reynolds number dependence of the relation between the solenoidal and irrotational 'components' of the nonlinearity as represented by $(\mathbf{u}\cdot\nabla)\mathbf{u}$ and $\boldsymbol{\omega}\times\mathbf{u}$, Fig. 8.6. There is a clear tendency of enhancement of solenoidality of the nonlinearity as the Reynolds number increases. Here too it is not clear what will happen when the Reynolds number will become very large.

The third example is the behavior of acceleration, see Fig. 8.7. Though the evidence is not conclusive, the indication is that there is no saturation of acceleration variance at large Reynold numbers either.

In Fig. 8.7 a number of points are shown obtained by Gulitskii et al. (2007b). The main feature is that there seems to be no saturation in the Re-dependence of the acceleration variance normalized on $\varepsilon^{3/2}\nu^{-1/2}$. This means that the scaling proposed by Yaglom (1949), see Monin and Yaglom (1975, p. 369) is not 'perfect' and the acceleration variance is larger than that proposed by Yaglom. The trend seen in Fig. 8.7 may be contaminated by the imperfections the method. The issue seems to be open and requires further far more precise measurements.

Another aspect of concern is that $a^2 \sim \varepsilon^{3/2}\nu^{-1/2} \sim u_0^{9/4}L^{-3/4}\nu^{1/4}$, i.e. the acceleration variance along with viscosity, ν, depends explicitly on the large scale characteristics, u_0, of the flow contrary to the claim that *the turbulent acceleration is determined largely by the very small-scale motions* $l \lesssim \eta$ (Monin and Yaglom 1971). Indeed, the observations show, see Fig. 7.3 right, that the conditional statistics of a^2 on u_0 show a significant statistical dependence, see also Biferale et al. (2004) and Lüthi et al. (2005) and references therein. The bottom line is that *fluid particle acceleration variance does not* (seem to) *obey K41 scaling at any Reynolds number* (Hill 2002).

Finally, as discussed above, the so called 'anomalous scaling' in the "inertial range" is due to viscous/diffusive effects with two options. One is just the finite Reynolds number effect, though the latest observations are at $Re_\lambda \sim 10^4$. The other one is due the influence of viscous/diffusive effects in the tails (i.e. strong events) of corresponding PDFs and is expected to be present at any Reynolds number. The evidence seems to favor the latter option. In this case there remains the question whether the "anomalous scaling" is Reynolds number independent.

These examples, along with other results, show that the issue of the asymptotic 'ultimate' regime/state of turbulent flows at very large Reynolds numbers remains open. Unfortunately, progress in mathematical treatment of the problem is very small if any, especially as related to basic physical aspects, for an overview see Sect. 10.2 in Tsinober (2009).

8.3 Symmetries

The starting point are the symmetries of the NSE and Euler

The Euler and the Navier Stokes equations are invariant under the following transformations, see e.g. Frisch (1995):

Translations in space (homogeneity) and time,
Full group of rotation including rotations and reflections (isotropy),
Galilean transformation $u(x,t) \Rightarrow u(x-Ut,t)+U$, $U = const$,
The Euler equation is in addition invariant under time reversal $t \Rightarrow -t$, $u \Rightarrow -u$, $p \Rightarrow p$.
scaling transformation $\mathbf{r} \Rightarrow \lambda\mathbf{r}$; $t \Rightarrow \lambda^{1-h}t$; $\mathbf{u} \Rightarrow \lambda^h\mathbf{u}$; $p \Rightarrow \lambda^{2h}p$, $\lambda > 0$ for any h.

The Navier–Stokes equations obey the scaling transformation for $h = -1$ only. However, it is not that uncommon belief that it "*may be justified at very high Reynolds number... that there are infinitely many scaling groups, labeled by their scaling exponent h, which can be any real number*", i.e. "*in the inviscid limit, the Navier–Stokes equation is invariant under infinitely many scaling groups, labeled by an arbitrary real scaling exponent h*" (Frisch 1995, pp. 18, 144), just like in the case of Euler equation. However, it is more than problematic why one can ignore the singular nature of the limit $Re \to \infty$ ($\nu \to 0$). Is not clear at all why one should forget that at any finite however large Re there is no such freedom as there is for a pure Euler case, especially when using experimental data at pretty moderate Reynolds numbers both for their "explanation" by the multifractal formalism and simultaneously its "confirmation" by the same data.

One of the issues of general nature is the belief//hypothesis that in "*turbulence at very high Reynolds numbers all or some possible symmetries are restored in a statistical sense* and that *for this it is necessary that flow should not be subject to any constraint, such as a strong large-scale shear, which would prevent it from 'accepting' all possible symmetries* and that *in order to achieve maximum symmetry it is advantageous not to have any boundaries. We could thus assume that the fluid fills all of the space* $\mathbb{R}^3$. *The unboundedness of the space does, however, lead to some mathematical difficulties. We shall therefore often assume periodic boundary conditions in the space variable* r (in a box of scale L) and recovering *the case of a fluid in the unbounded space* $\mathbb{R}^3$ *by letting* $L \to \infty$ (Frisch 1995, pp. 11, 14).

In this context the first question is how relevant is such an idealized turbulent flow to any real one even at the qualitative level and the second question is about how real, even approximately is the just mentioned hypothetical(!) restoring of the symmetries in the statistical sense . The experimental and DNS evidence for this hypothesis is scarce, very limited and at best marginal.

It is for the above reasons that Kolmogorov (1941a) proposed a hypothesis postulating that all the symmetries (including homogeneity and isotropy) of the Navier–Stokes equations are restored in the statistical sense in **any** turbulent flow for large enough Reynolds number *in sufficiently small regions G of the four-dimensional space* (x_1, x_2, x_3, t) (i.e. time-space) *not lying close to the boundaries of the flow or its other special regions*[5]—except for the one involving scaling. In order to handle this latter Kolmogorov introduced the concept of inertial range (IR) with the assumption that the statistical properties of this conventionally defined IR (CDIR) are independent of viscosity, thus assuming the scale invariance in the IR with the mean dissipation $\langle\epsilon\rangle$ being the only governing parameter.

There are numerous interpretations and ascribings of many things not belonging to Kolmogorov. The relevant one is about assumption is localness of interactions (Frisch 1995, p. 104)—we use the term locality. The above involves some decomposition just like the suggestion by Kraichnan (1974) about the 'proper inertial-range quantity: the local rate of energy transfer' with all the problems in proper definitions of this quantity and use of Fourier transform. One more issue deserving special mentioning is the claim that the Euler equations are valid in the inertial range. For example (among several), *In the inertial range we can neglect the viscosity and forcing and study the Eulerian dynamics of an ideal incompressible fluid* (Migdal 1995). Even if the inertial range does exist (which is not) after excluding some range

[5]Frisch (1995) presents this in the form of his hypothesis H1 (p. 74), but omits to mention that it is due to Kolmogorov: there is no presentation of the hypothesis of local isotropy in his book. A similar omitting is pretty ubiquitous, the latest example being (Falkovich 2009).

of scales such a statement by itself does not seem to make any sense—after all Euler is a differential equation. Finally, Kolmogorov used exclusively dimensional analysis and similarity arguments and did not make any claims about things like validity of Euler in the inertial range.

Anyway, it is a common view in consensus that in real systems the symmetries are broken due to instabilities, forcing (excitation), inflow, initial and boundary conditions, but that following Kolmogorov the symmetries are expected to be restored locally in statistical sense at high Reynolds numbers. However, this expectation ignores the role of nonlocal nature of turbulence. The consequence due of effects of nonlocality is that, generally, one cannot apply results for globally homogeneous, isotropic turbulent flows to flows which are approximately homogeneous in a bounded region of flows and otherwise are non-homogeneous, because the nonlocality makes local homogeneity, isotropy, etc. impossible unless the whole (infinite extent) flow is such, which is trivially impossible. The "quasi-homogeneous" behavior of some quantities in bounded regions is misleading due to several factors, the main being the nonlocal nature of turbulence. Typically these are quantities and relations exhibiting "quasi-Gaussian" manifestations. On this see Sect. 6.8.2, Chap. 6 in Tsinober (2009).

As mentioned the attraction of Kolmogorov hypotheses is that they concern the local properties of any turbulence, not necessarily homogeneous or isotropic or decaying or stationary, provided the Reynolds number is large enough. The consistency of these hypotheses is debated for quite a while both from the theoretical point of view (Gkioulekas 2007; Hill 2006 and references therein) and also experimentally, but as with other similar issues without much progress from the fundamental point. One of the main problems is the inherent property of nonlocality questioning the validity of the above hypotheses on local homogeneity/isotropy in bounded flow domains surrounded by nonhomogenous/nonisotorpic flow regions.

A similar controversial situation is with spatially developing flows claimed to possess the property of self-similarity in analogy with von Karman (1943) and von Karman and Howarth (1938) as an attempt to "solve" their famous (KH) equation, containing both second and triple order correlations. This required a hypothesis to close the KH equation which is known as the so-called self-preservation (or self-similarity) hypothesis. Such a claim is essentially an assumption of locality both in time and space: the turbulent motion at some point in time and space is assumed to be defined by its immediate proximity.

A special example is about reflectional symmetry, e.g. involving helicity defined as the integral of the scalar product of velocity and vorticity, $H = \int \mathbf{u} \cdot \boldsymbol{\omega} dx$, see Tsinober (2004). The hypothesis of local isotropy (K41a) includes restoring of all the symmetries in small scales. Thus one expects restoring of reflection-invariance at small scales. However, to maintain finite "helicity dissipation" to balance the finite helicity input in a statistically stationary turbulence helically forced at large scales the tendency to restore reflection symmetry at small scales can not be realized. This is because helicity dissipation is associated with broken reflection symmetry at small scales, since helicity dissipation, $D_H = -\nu H_s$ is just proportional to the superhelicity $H_s = \int \boldsymbol{\omega} \cdot \text{curl}\, \boldsymbol{\omega} dx$, showing the lack of reflection symmetry of the small scales.

The important point is that helicity dissipation is vanishing if reflectional symmetry holds in small scales. Moreover, if the helicity dissipation should remain finite as the Reynolds number increases this lack of reflectional symmetry should increase since the dissipation of helicity ($D_H = -\nu H_s$) is proportional to viscosity. A noteworthy phenomenon is breaking of reflectional symmetry in a turbulent flow with initially vanishing pointwise helicity density $\mathbf{u} \cdot \boldsymbol{\omega} \equiv 0$ (Shtilman et al. 1988)—a result that can be interpreted in terms of statistical instability of turbulent flow with reflectional symmetry to disturbances breaking this symmetry, see also a non-trivial experiment in Kholmyansky et al. (2001c).

To repeat, the role of nonlocality is the same in the issue concerning the scale invariance in the conventionally defined inertial range (CDIR). The experimental results at high Reynolds numbers show unequivocally that such an object is ill defined and that the CDIR in reality contains dissipative events which among other things are responsible for what is called anomalous scaling in the CDIR.

The bottom line is that there is no answer to the question by Kraichnan (1974): *How a theoretical attack on the inertial-range problem should proceed is far from clear,* because this question relates to a nonexistent object. A similar statement is true as concerns the 'theories' such as multifractal formalism (Frisch 1995), see also She and Zhang (2009), attempting to 'explain' intermittency and anomalous scaling in such a nonexistent object, for more see Chap. 5, Chaps. 6 and 9 in Tsinober (2009).

On Analogies Involving Symmetries and Related Issues A general aspect is that it is much less than sufficient for two systems to share the same basic symmetries, conservation laws and some other general properties, to have the same or similar basic properties. However, it is a common practice and multitude of models such as, e.g. using results from the so called "passive turbulence" (i.e. a passive scalar in a random velocity field—not necessarily physical) to draw conclusions about genuine turbulence obeying the NSE. However, this is not really the case, as, e.g. pointed out by Kraichnan (1974) in a counter example of a '*dynamical equation is exhibited which has the same essential invariances, symmetries, dimensionality and equilibrium statistical ensembles as the Navier–Stokes equations, but which has radically different inertial-range behavior.*' The issue is much broader. Using the above properties only many essential properties are missed, such as a variety of dynamically relevant geometrical aspects. The first example, concerns the interaction of vorticity and strain and enstrophy production. Here of utmost importance are the geometrical relations between the vorticity and the eigenframe of the rate of strain tensor, see Chap. 6 and Sect. 6.4 in Tsinober (2009).

As described in Sect. 5.1.1 in Tsinober (2009) the issue extends into broad misinterpretations of a variety of analogies such as between the genuine (e.g. NSE) turbulence and passive "turbulence", i.e. evolution of passive objects in random (or just not too simple) velocity fields. The differences are more than essential, though there are numerous claims for *the well-established phenomenological parallels between the statistical description of mixing and fluid turbulence itself* (Shraiman and Siggia 2000). Of special interest here is the anomalous scaling in genuine versus passive turbulence, e.g. a recent statement by Eyink and Frisch (2011, p. 362): *...Kraichnan's model of a passive scalar advected by a white-in-time Gaussian random velocity has become a paradigm for turbulence intermittency and anomalous scaling*. The authors mean the genuine NSE turbulence. We just repeat that the analogy between genuine and passive turbulence is illusive and mostly misleading. Far more can be found on misuse and misinterpretations of analogies in Chap. 9 in Tsinober (2009) and references therein.

On Periodic Boundary Conditions There is considerable evidence that the statistical properties of some turbulent flows with the same geometry at very modest Reynolds numbers are invariant of the boundary and even initial conditions (BC and IC). For example, typical DNS computations of NSE of turbulent flows (e.g., in a circular pipe and a plane channel, in a cubic box, etc.) involve extensive use of periodic BC. The results of these computations agree well with those obtained in laboratory experiments, in which the BC have nothing to do with periodicity. Indeed, the correlation coefficient between two values of any quantity at the opposite ends of such boundaries (i.e., the points separated at maximal distance in the flow domain) is precisely equal to unity and close to unity for the points in the proximity of such boundaries, whereas in any real flow the correlation coefficient becomes very small for points separated by a distance on the order of (and larger than) the integral scale of turbulent flows.) and in which the IC were totally different from those in DNS. No explanation of this kind of invariance is known so far, but it is natural to expect that it is related to some kind of hidden symmetry(ies) of the NSE. If such exist, this may be the reason for the similarity of results obtained via DNS of NSE in, e.g., periodic boxes by various forcing (different deterministic, random/stochastic, etc.).

The above property, however, is not universal and there are many examples of long memory of turbulent flows which do 'remember', e.g., the inflow and initial conditions. As mentioned the reasons for this behavior may involve nonlocal properties in conjunction with the changing environment so that this "memory" is an attribute of a transient phenomenon in spatially developing flows due to continuously changing conditions preventing approach to some universal state/behavior.

8.4 Universality

The issue of universality is one of several continuously debated controversies in the problem of turbulence. This includes the meaning of the term 'universality' which always refers to behavior at large Reynolds numbers. For example, one issue involves the invariance of Reynolds number of (some) properties of a particular turbulent flow at large enough Reynolds numbers (here and then). Another issue is concerned with the universality of (mostly scaling) properties of small-scale turbulence assumed to be the same in any flow, which has remained for more than fifty years one of the most active fields of inquiry. Derivation of scaling properties of fully developed turbulent flows directly from the Navier–Stokes equations analytically is one of the most popular illusive goals of theoretical research. Since the Kolmogorov papers (1941a, 1941b), there exists still almost a religious belief in some universal properties of small scale turbulence.[6] This belief was strengthened by achievements in dynamical chaos, such as the discovery of some universal numbers by Feigenbaum, etc.

The expectation of universality in turbulence stems from the nonlinear nature of turbulence. However, this expectation is to a large extent did not come true due to much neglected set of properties of essentially nonlocal nature. This includes practically all cases where quantitative universality was expected. In other words, the problems with universality are due to the competition between nonlinearity and non-locality, the latter is also aided by linear processes, such as in the proximity of walls (Kim 2012 and references therein). However, as it stands now, the main problems are with quantitative universality, i.e. universality of numbers and the necessity to distinguish between quantitative universality and the qualitative one, i.e. involving some more general universal features and processes.

8.4.1 Quantitative Universality

With the exception of the 2/3 (5/3) and the 4/5 laws there appeared to exist no quantitative universality so far: the first doubt came from the famous remark by Landau in the first Russian edition of Fluid Mechanics by Landau and Lifshitz about the fluctuations of energy dissipation rate. These were followed by various 'universal' corrections, which did not appear to be universal either. These corrections were followed by the (multi) fractal approach using either the so-called $D(h)$ or $f(\alpha)$ formalisms, in which the *functions* $D(h)$ and/or $f(\alpha)$ are assumed to be universal. However, they do not seem to be universal either. There is quite solid evidence accumulated during the last fifty years against the most beautiful hypothesis on the restoring of the symmetries in the statistical sense of the Navier–Stokes equations locally in time and space, i.e. local isotropy together with scale invariance. And so people started to look for some universality in the anisotropic properties of turbulent flows, see references in Biferale and Procaccia (2005). This involves the SO(3) decomposition of tensorial objects again assuming universal scaling behavior in r of each component of the decomposition. Consequently there is no "simple" scaling of, say, structure functions in r, but rather the different terms of the SO(3) decomposition each with its own scaling exponent assumed to be universal. Thus all the attraction of simple scaling as in Kolmogorov (1941a, 1941b) has gone.

The assumption of universality has no serious justification and is more a kind of a belief much weaker than the belief in the inertial range. Moreover, it is not clear at all

[6] That far that *one of the principal incentives for writing this book was a desire to summarize the development of the idea of a universal local structure in any turbulent flow for sufficiently large Reynolds number* (Monin and Yaglom 1971, p. 21). This is 1600 pages total.

why each "sector" of the irreducible representation is expected to have its own universal scaling exponent independently of the physical/dynamical nature/underlying mechanisms of anisotropy such as mean shear, strain, rotation, stratification (both stable and unstable), magnetic field, etc. The expectation of universality is especially problematic in case of strong anisotropy (Q2D) in all the above cases. There is a claim that the amplitudes of the various contributions are nonuniversal and that it is possible to fit the experimental data by keeping fixed the scaling properties and adjusting only the prefactors. It is also noteworthy that in determination of anisotropic scaling exponents one encounters the same kind of difficulties as those known from previous experience. This however, does not prove much regarding universality and may well be at best the "right result not necessarily for the right reason". One more difficulty may arise due to nonuniqueness of the SO(3) decomposition in the sense that there exists more than one possibility to chose its basis in case when the SO(3) decomposition is applied to tensorial objects. There are also similar claims on universality related to passive objects. This kind of a claim is quite surprising, as passive objects are governed by linear equations and thus its statistics and scaling exponents are expected to be sensitive to the statistics of the velocity field.

The nonuniversality of small scale structure of turbulence is mostly due to nonlocality along with the difference in the mechanisms of large-scale production. The latter depend on the particularities of a given flow and thus are not universal. The large scales, especially their edges seem to be responsible for the contamination of the small scales. This 'contamination' is unavoidable even in homogeneous and isotropic turbulence, since there are many ways to produce such a flow, e.g., many ways to produce the large scales. The situation is complicated by the reaction back of the small scales on the large ones.

On the Special Status of Scaling Exponents of the Second and Third Order Structure Functions The empirically observed Reynolds number independent behavior of some global characteristics of turbulent flows at large *Re*, such as the total dissipation, the drag of bluff bodies and resistance in many other configurations, comprises one of the quantitative universal properties of turbulence at large *Re*. This is distinct from some possibly universal (mostly scaling) properties of small-scale turbulence. The exception in some sense are the scaling exponents of the second and third order structure functions.

It is widely thought that the Kolmogorov similarity hypotheses imply that the mean dissipation, remains finite/non-vanishing as $Re \to \infty$, whereas they are based on an assumption that this is the case. The issue is whether the properly normalized mean dissipation (usually by U^3/L^{-1} as suggested by Taylor 1935) tends really to a finite limit as $Re \to \infty$ ($\nu \to 0$), or is *Re* dependent even at very large Reynolds numbers. There is much speculation about this subject, while the experimental and recent evidence from DNS favoring the former is solid but still limited, though in a variety of fluid engineering practice this fact was recognized long ago in a great variety of flow configurations since the end the 19th century. Since the finite limit of the mean dissipation at large *Re* defines a unique scaling exponent in the 2/3 law and since the 4/5 law is a consequence of the Navier–Stokes equations, the two scaling exponents, $\zeta_2 = 2/3$ and $\zeta_3 = 1$, possess a special status provided that the inertial range is a well defined concept (which is not) and the mean dissipation is indeed Reynolds number independent as $Re \to \infty$. It is noteworthy that rigorous upper bounds of the long time averages of dissipation are independent of the Reynolds number at large Re (Doering 2009 and references therein), and thereby are consistent with the experimental results. This, however, does not fully resolve the problem since there are no results on the lower bounds, except trivial values corresponding to the laminar flows. There are also experimental results on the energy spectra at large Reynolds numbers with the exponent steeper than $-k^{-5/3}$ (see references and new results in Tsinober 2009) and there is a need for different Reynolds numbers in the experimental confirmations of the 4/5 law at large Reynolds number (Kholmyansky and Tsinober 2009). Thus it makes sense to keep in mind the possible alternatives, see e.g. Grossman (1995), Long (2003) and references in Abe and Antonia (2011).

In other words, it seems that the dream of quantitative universality of turbulence, i.e. universality of numbers, may never come true.[7] The main reason is the nonlocality leading not only to breaking of the symmetries embodied in the Navier–Stokes equations, but preventing them to be restored in the statistical sense at high Reynolds numbers, along with the ill posedness of the concept of inertial range as discussed below in the next subsection on qualitative universality.

8.4.2 *Qualitative Universality*

Though there may not exist such a thing like quantitative universality of turbulence, i.e. universality of numbers, there seems to exist a qualitative one. The concept of qualitative universality is not just a fuzzy idea. First, the major qualitative features of turbulent flows as described in Chap. 1 are universal for all turbulent flows arising in qualitatively different ways and circumstances and generally characterize turbulent flows as a whole. It is these qualitative features which do not depend on the diverse processes/ways by which flows become turbulent. Second, these qualitative features possess a number of quantitative attributes/features which are more specific for turbulent flows and are of special interest. Third, as mentioned it is natural to use the major universal qualitative manifestations of turbulent flows as a basis and as a first step to 'define' what is turbulence.

In other words, apart of general qualitative properties of turbulence as apparent randomness, enhanced effective diffusivity and dissipation, rotational nature, and others as discussed in Chaps. 1 and 6, of particular interest are more specific qualitative universal properties of turbulent flows along with the quantitative attributes of the former.

Below we attend a number of most important issues on and/or related to qualitative universality.

[7]For other negative statements about universality see, for example, Saffman (1978) and Hunt and Carruthers (1990).

Turbulent flows possess (empirically) stable statistical properties (SSP), not just averages but almost all statistical properties. In case of statistically stationary flows the existence of SSP seems to be an indication of the existence of what mathematicians call attractors. But matters are more complicated as many statistical properties of time-dependent in the statistical sense turbulent flows possessing no attractor but stable SSP are quite similar at least qualitatively to those of statistically stationary ones as long as the Reynolds number of the former is not too small at any particular time moment of interest. This can be qualified as qualitative temporal universality/memory.

Velocity Derivatives, Self Amplification, Tennekes and Lumley Balance The most important is the process of self-amplification of velocity derivatives, both strain and vorticity which turns turbulence into a strongly dissipative and rotational phenomenon. These two concomitant processes posses a quantitative universal aspect as reflected in the Tennekes an Lumley balance. It should be stressed again that there is an essential and qualitative difference between the process of self-amplification of strain and other similar processes. It is a specific feature of the dynamics of turbulence having no counterpart in the behavior of passive objects. In contrast, the process of self-amplification of vorticity, along with essential differences, has a number of common features with analogous processes in passive vectors; in both the main factor is their interaction with strain, whereas the production of strain is a self-production provided that the flow is rotational.

The amplification of derivatives in other systems as passive objects is a property which can be seen as universal not only for any random fluid flow, be it real or artificial, such as the Gaussian velocity field, but of any fluid flows that are Lagrangian chaotic, many of which are simple laminar in the Eulerian setting.

On Universal Aspects of Turbulence Structure In dynamical systems, one looks for structure in the *phase space* (Shlesinger 2000; Zaslavsky 1999), since it is relatively 'easy' due to low dimensional nature of the problems involved. In turbulence nothing is known about its properties in the corresponding infinite or very high dimensional phase space.[8] Therefore, it is common to look for structure in the *physical space* with the hope that the structure(s) of turbulence—as we observe it in *physical space*—is (are) the manifestation of the generic structural properties of mathematical objects *in phase space*, which are called attractors and which are invariant in some sense. In other words, the structure(s) is (are) assumed to be 'built in' in the turbulence independently of its (their) origin—hence universality. However, as mentioned, the expectation for universal numbers seems to be unjustified. It is more natural to expect universal qualitative statistical features in the physical space rather than universal numbers. Indeed, some of such features have been already observed, which are common for very different—essentially all known—turbulent flows. These are not only the general qualitative features of turbulent flows as described in Chap. 1, but rather specific ones.

---

[8] Hopf (1948) conjectured that the underlying attractor is finite dimensional due to presence of viscosity.

We bring three examples with features which are essentially the same for all known incompressible flows such as grid turbulent flow, periodic flow in a computational box, turbulent boundary layer and channel flow, mixing layer and compressible flows as well. Such features can be seen as universal statistical manifestations of the structure of turbulent flows.

The first example, is the so-called 'tearing drop' feature observed in the invariant map of the second invariant, $Q = \frac{1}{4}(\omega^2 - 2s_{ik}s_{ik})$, versus the third invariant $R = -\frac{1}{3}(s_{ik}s_{km}s_{mi} + \frac{3}{4}\omega_i\omega_k s_{ik})$ of the velocity gradient tensor $\partial u_i/\partial x_k$, such as shown in Fig. 6.1. This feature appears to be essentially the same for all known turbulent flows. We draw the attention to the 'tail' of the teardrop which is mainly located in the quadrant $Q < 0, R > 0$, in which most of turbulent activity happens in a variety of ways. The important point is that this is the region dominated by strain as compared with enstrophy, $2s_{ik}s_{ik} > \omega^2$, and by *production* of strain as compared with production of enstrophy, $-s_{ik}s_{km}s_{mi} > \frac{3}{4}\omega_i\omega_k s_{ik}$. This is in full conformity with the behavior of nonlinearities in these regions, see Sect. 6.5 and Chap. 10.1.3 in Tsinober (2009).

The second example is related to depression of nonlinearity. We mention here one aspect of this problem, which seems to be universal in the sense that it is true for different flows (including shear) and different Reynolds numbers, though the evidence is still quite limited. Namely, all key nonlinearities appear to be much stronger in the strain dominated regions rather than in regions with concentrated vorticity, in contrast to the common expectation that, for example, the vorticity amplification process will be strongest where the vorticity already happens to be large. The regions with concentrated vorticity are in approximate equilibrium in the sense that the rate of enstrophy production is in approximate balance with the viscous destruction in these regions even at low Reynolds numbers. Therefore, their life time is considerably larger than the life time of the regions dominated by strain, which are in strong disequilibrium in the sense that the rate enstrophy production is much larger than its destruction by viscosity in these regions. This is reflected in the R–Q diagram as the strongest activity in the fourth quadrant in which Q and R are negative, i.e. $2s^2 > \omega^2$ and $-s_{ik}s_{km}s_{mi} > \frac{3}{4}\omega_i\omega_k s_{ik}$.

It is noteworthy that the issue of depression of nonlinearity was initiated via comparison of nonlinearities in real turbulent flows with their Gaussian counterparts with the results that turbulent flows tend to deplete nonlinearity significantly. The problem is that such a comparison is too narrow as it is meaningful for even moments only, because for a Gaussian field all odd moments such as, e.g. strain and enstrophy production, vanish identically. In other words, measured by odd moments the real nonlinearity is infinitely larger, see Tsinober (2009, p. 154).

The third example is related to geometrical statistics for example of enstrophy production as a process of interaction of vorticity and strain. These are various alignments between vorticity and the eigenbasis of the rate of strain tensor, and between vorticity and the vortex stretching vector. It appears that these and many other similar properties are the same for all known flows and, moreover, for a broad range of Reynolds numbers. For example, the character of the above alignments is essentially the same at $Re_\lambda \sim 10^2$ and $Re_\lambda \sim 10^4$.

The same similarity is observed recently for a variety of alignments and other properties associated with fluid particle accelerations (Gulitskii et al. 2007b) and for passive scalars (Gulitskii et al. 2007c) such as alignments of passive scalar gradient, $\mathbf{G}$ with the eigenframe $\boldsymbol{\lambda}_i$ of the rate of strain tensor, conditional averages on ω^2 and s^2 of the production $-G_i G_j s_{ij}$, tilting of $\mathbf{G}$ and some others.

There is a far more difficult issue of universality of finite objects in turbulence, which is discussed in the next chapter.

Chapter 9
Intermittency and Structure(s) of and/in Turbulence

Abstract Intermittency specifically in genuine fluid turbulence is associated mostly with some aspects of its spatiotemporal structure. Hence, the close relation between the origin(s) and meaning of intermittency and structure of turbulence. Just like there is no general agreement on the origin and meaning of the former, there is no consensus regarding what are the origin(s) and what turbulence structure(s) really mean. At the present state of matters both issues are pretty speculative and an example of 'ephemeral' collection of such is given in this chapter. We have to admit at this stage that structure(s) is(are) just an inherent property of turbulence.

Structureless turbulence is meaningless. There is no turbulence without structure(s). Every part (just as the whole) of the turbulent field—including the so-called 'structureless background'—possess structure. Structureless turbulence (or any of its part) contradicts both the experimental evidence and the Navier–Stokes equations. The claims for 'structureless background' is a reflection of our inability to 'see' more intricate aspects of turbulence structure: intricacy, complexity and 'randomness' are not synonymous for absence of structure.

What is definite is that turbulent flows have lots of structure(s). The term structure(s) is used here deliberately in order to emphasize the duality (or even multiplicity) of the meaning of the underlying problem. The first is about how turbulence 'looks'. The second implies the existence of some entities. Objective treatment of both requires use of some statistical methods. It is thought that these methods alone may be insufficient to cope with the problem, but so far no satisfactory solution was found. One (but not the only) reason—as mentioned—is that it is not so clear what one is looking for: the objects seem to be still elusive. For example, there is still a non-negligible set of people in the community that are in a great doubt that the concept of coherent structure is much different from the Emperors's new Clothes.

An example of acute difficulties described in this chapter is associated with high dimension of what is called structures so that simple single parameter thresholding is inappropriate to make on them statistics due to the painful question how really "similar" are all these if the individual members of such an ensemble are defined by one parameter only. The view that turbulence structure(s) is(are) simple in some sense and that turbulence can be represented as a collection of simple objects only seems to be a nice illusion which, unfortunately, has little to do with reality. It seems somewhat wishfully naive to expect that such a complicated phenomenon like tur-

A. Tsinober, *The Essence of Turbulence as a Physical Phenomenon*,
DOI 10.1007/978-94-007-7180-2_9,

bulence can merely be described in terms of collections of only such 'simple' and weakly interacting object.

The nature and characterization of the structure(s) of turbulent flows are among the most controversial issues in turbulence research with extreme views on many aspects of the problem—in words of Richard Feynmann (1963, pp. 41–12), *holding strong opinions either way*. For example, it is common in the vast literature on turbulence to consider the terms *statistical* and *structural* (and also *deterministic*) as incompatible or even contradictory, see for example Dwoyer et al. (1985), Lumley (1989). However, there are common points as well. For instance, it is mostly agreed that turbulence definitely possesses structure(s) (whatever this means) and that intermittency, which is addressed in the following subsection, is intimately related to *some* aspects of the structure(s) of turbulence.

At present it became clear that it is a misconception to contrapose the *statistical*, the *structural* and the *deterministic* and that they represent different facets/aspects of the same problem, so that there is no real gap between structure(s), statistics (but not in the sense of absence of laws) and determinism. Just like it seems impossible to separate the structure(s) from the so called 'random structureless background' or the '*random processes from the nonrandom processes*', Dryden (1948) due to strong interaction and nonlocality, both between individual structures, and between structures and the 'background'. In other words, there is no turbulence *without* structure, every part of the turbulent field just like the whole possess structure.[1] Structureless turbulence or any its part contradicts both the experimental evidence and the Navier–Stokes equations. It is noteworthy that the statement that turbulence has structure is in a sense trivial: *to say that turbulent flow is 'completely random' would define turbulence out of existence* (Tritton 1988, p. 295).

One of the basic properties underlying both aspects, i.e. intermittency and structure(s) is the essentially non-Gaussian nature of turbulence, which follows from NSE, e.g. Novikov (1967), Lumley (1970), and there are numerous experimental observations on non-Gaussianity of turbulence, see e.g. Sect. 6.8 in Tsinober (2009) on non-Gaussian nature of turbulence, e.g. the outstanding non-zero odd moments: the strain and enstrophy production $-\langle s_{ij}s_{jk}s_{ki}\rangle$, $\langle \omega_i\omega_j s_{ij}\rangle$. It has to be stressed that (i) non-Gaussianity and intermittency/structure are not synonymous as not any non-Gaussian field is turbulent and (ii) non-Gaussianity is only a statistical manifestation of intermittency/structure.

As a simple illustration one can see that even if the flow field is initially Gaussian, the dynamics of turbulence makes it non-Gaussian with finite rate. This is seen by looking $\langle \ldots \rangle$ at the equation for $\langle \omega_i\omega_j s_{ij}\rangle$ (dropping the viscous term).

$$D\langle \omega_i\omega_j s_{ij}\rangle / Dt = \langle \omega_j s_{ij}\omega_k s_{ik}\rangle - \langle \omega_i\omega_j \Pi_{ij}\rangle .$$

[1] There are proposals to *scan out* the structure(s). In fact there is no way to do so, since *structure* is everywhere. Even the so-called 'simple' structures, worms, sheets, etc. are 'renormalized' by the background.

For a Gaussian velocity field $\langle \omega_i \omega_j s_{ij} \rangle_G = 0$, $\langle \omega_i \omega_j \Pi_{ij} \rangle_G = 0$ and $\langle \omega_j s_{ij} \omega_k s_{ik} \rangle_G = \frac{1}{6}\langle \omega^2 \rangle^2 > 0$. Since the quantity $\omega_j s_{ij} \omega_k s_{ik} \equiv W^2$, $W_i = \omega_j s_{ij}$, it is positive pointwise for any vector field. Hence for an initially Gaussian field

$$\left\{ D\langle \omega_i \omega_j s_{ij} \rangle / Dt \right\}_{t=0} = \left\{ \langle \omega_j s_{ij} \omega_k s_{ik} \rangle \right\}_{t=0} > 0.$$

It follows that, at least for a short time interval t, the mean enstrophy production will become positive.

The above equations and similar ones for $\langle s_{ij} s_{jk} s_{ki} \rangle$ can be seen as one of the manifestations of the statistical irreversibility of turbulent flows (Betchov 1974; Novikov 1974). The corresponding dynamical instantaneous (inviscid) equations are reversible. Hence, the term "statistical". One would claim that the Kolmogorov 4/5 law in the conventionally defined inertial range belongs to the same category, but as described above it appears to be not a purely inertial relation. Another aspect of irreversibility is related to the dissipative nature of turbulence—its not "slightly" dissipative at whatever large Reynolds number.

Turbulence—being essentially non-Gaussian—is such a rich phenomenon that it can 'afford' a number of Gaussian-like manifestations, some of which are not obvious and even nontrivial, for examples see Sect. 6.8.2 in Tsinober (2009). Hence the importance of parameters related to the non-Gaussian nature of turbulence and not only in various contexts of intermittency and structure(s).

9.1 Intermittency

At any instant the production of small scales is... occurring vigorously in some places and only weakly in the others (Tritton 1988).

Intermittency is a phenomenon where Nature spends little time, but acts vigorously (Betchov 1993).

Typical distribution of scalar and vector fields is one in which there appear characteristic structures accompanied by high peaks or spikes with large intensity and small duration or spatial extent. The intervals between the spikes are characterized by small intensity and large extent (Zeldovich et al. 1990).

The term intermittency is used in two distinct (but not independent) aspects of turbulent flows. The first one is the so-called external intermittency. It is associated with what is called here partly turbulent flows, specifically with the strongly irregular and convoluted structure and random movement of the 'boundary' between the turbulent and nonturbulent fluid.

The second aspect is the so-called small scale, internal or intrinsic intermittency and is associate with spotty temporal and spatial patterns of the small scale structure(s) in the interior of turbulent flows.

9.1.1 The External Intermittency and Entrainment

This kind of intermittency was studied first by Townsend (1948) and involves the so called entrainment which is one of the most basic processes of transition from laminar to turbulent state with the coexistence of both in one flow. It is associated with the coexistence of laminar and turbulent flow regions—an observer located in the proximity of either side of the 'mean' boundary between these regions observes intermittently laminar and turbulent flow in the form of a signal similar to that as in Fig. 21.4 in Tritton (1988), clearly demonstrating the external intermittency of in the wake past a circular cylinder. Here we again encounter the question about what turbulent is. Looking at signals like the mentioned one clearly sees what is turbulent and what is laminar. But the question is how one can say whether a small part of flow is turbulent. In other words if turbulence is to be identified by statistical means, then what is the meaning of 'turbulent' locally? This involves taking decisions about what is turbulent using some conditional criterion, see discussion and references in Kuznetsov et al. (1992).

The main mechanism by which nonturbulent fluid becomes turbulent locally as it crosses the interface is due viscous diffusion of vorticity across the interface. As this process is associated with small scales it is thought to be the reason why the interface appears sharp compared to the scale of the whole flow.

However, at large Reynolds numbers, the entrainment rate and the propagation velocity of the interface relative to the fluid are known to be independent of viscosity. Therefore *the slow process of diffusion into the ambient fluid must be accelerated by interaction with velocity fields of eddies of all sizes, from viscous eddies to the energy-containing eddies so that the overall rate of entrainment is set by large-scale parameters of the flow* (Townsend 1976). That is *although the spreading is brought about by small eddies* [viscosity] *its rate is governed by the larger eddies. The total area of the interface, over which the spreading is occurring at any instant, is determined by these larger eddies* (Tritton 1988). This is analogous to independence of dissipation of viscosity in turbulent flows at large Reynolds numbers. In other words, small scales do the 'work', but the amount of work is fixed by the large scales in such a way that the outcome is independent of viscosity. This shows that independence of some parameter of viscosity at large Reynolds numbers does not mean that viscosity is unimportant. It means *only* that the (cumulative) effect of viscosity is Reynolds number independent, for more see Tsinober (2009), Wolf et al. (2012a, 2012b) and references therein.

9.1.2 The Small Scale, Internal or Intrinsic Intermittency

The second aspect is the so-called small scale, internal or intrinsic intermittency. It is usually associated with the tendency to spatial and temporal localization of the 'fine' or small scale structure(s) of turbulent flows with large probability of taking values of some variable both very large and very small compared to its standard deviation. An important feature is that regions of exceptionally high values

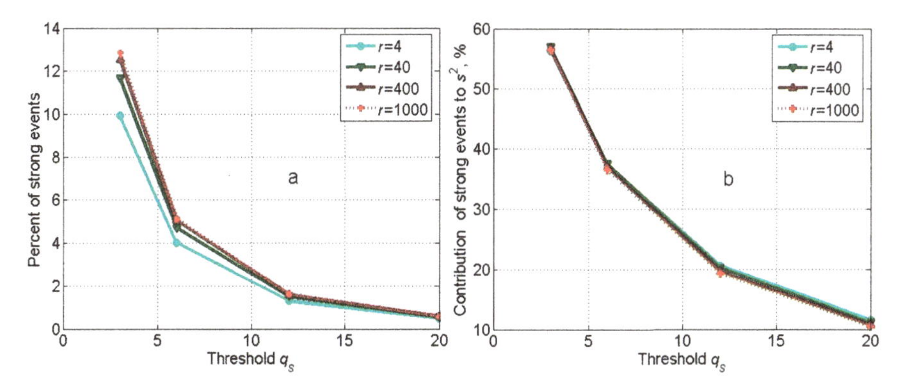

Fig. 9.1 Percent of the strong dissipative events as defined in the text (**a**) and their contribution to the total dissipation (**b**) as a function of the threshold q for various separations r. The value of separation is not relevant as concerns the contribution of the dissipative events to the total dissipation (Kholmyansky and Tsinober 2009)

make disproportionately large contribution (sometimes even dominant) to the integral properties even though occupying a small fraction of the flow. Consequently, regions with "voids" (low intensity) occupy a "disproportionately" large fraction of the flow and thus being statistically dominating, but contribute far less to the dynamics. For example, in field experiments with $Re_\lambda \approx 10^4$ the dissipative events with dissipation exceeding the mean 12 times (i.e. $\epsilon < 12\langle\epsilon\rangle$) contribute to $\langle\epsilon\rangle$ about 20 % while taking about 1 % of the total volume, see Fig. 9.1. The instantaneous values of dissipation exceed $10^4\langle\epsilon\rangle$.

Similar behavior is observed for other variables exhibiting intermittency, e.g. enstrophy and enstrophy and strain production, see Tsinober (2009).

Our concern here is with the intermittency of this second kind. It is noteworthy that in a broad sense intermittency is a ubiquitous phenomenon occurring in a great variety of qualitatively different systems, see Zeldovich et al. (1990) for a lively exposition of a wide number of different systems exhibiting intermittency, and also Vassilicos (2001). The main common features of all of them is (space/time) irregularity and localization (both spatial and temporal) of their 'fine' structure. However, this is not enough to define intermittency. For example, almost any nonlinear function or almost any nonlinear functional of a random Gaussian field is intermittent in the above sense, though random Gaussian fields by definition lack any intermittency.

We do not make a special distinction between the intermittency in the inertial and dissipative ranges as both are not well defined.

The intermittent nature of the small scale structure of turbulent flows was foreseen by Taylor (1938b): ...*the view frequently put forward by the author that the dissipation of energy is due chiefly to the formation of very small regions where vor-*

ticity is very high.[2] Indeed, *such behavior is not unexpected since the viscous term in the NSE contains the highest-order derivative while high Reynolds number turbulence involves the limit* $\nu \to 0$. *This limit is a singular perturbation problem and localized regions in which gradients are large should be expected to form* (Orszag 1977). Along with nonlocality this means that the consequence of intermittency is not just a small "correction" in the properties of turbulence with the possible exception of the quasi-Gaussian ones.

The phenomenon of small scale intermittency was observed by Batchelor and Townsend (1949) in experiments with turbulent grid flows and in a wake past a circular cylinder.

Batchelor and Townsend (1949) wrote: *The basic observation which requires explanation is that activation of large wave-numbers is very unevenly distributed in space. These space variations in activation can be described as fluctuations in the spectrum at large wave-number... As the wave-number is increased the fluctuations seem to tend to an approximate on-off, or intermittent variation. Whatever the reason for the occurrence of these fluctuations, they appear to be intrinsic to the equilibrium range of wave-numbers. All the evidence is consistent with the inference that the fluctuations are small in the region of smallest wave-numbers of equilibrium range and become increasingly large at larger wave-numbers... the mean separation of the visible activated regions is comparable with the integral scale of the turbulence, i.e. with the size of the energy-containing eddies.* Batchelor and Townsend (1949, pp. 252–253) obtained some evidence that the deviation from Gaussianity is stronger as the Reynolds number is increased. However, they did not appreciate this effect and claimed that the flatness factors seems to vary a little with Reynolds number, though this factor changed from 5 to 7 for the third order derivative; see their Fig. 5 for the flatness factor of velocity derivatives of different orders and different Reynolds numbers. This was confirmed by a number of subsequent experiments such as by Kuo and Corrsin (1971), for an updated overview of the subsequent results see, e.g. Sreenivasan and Antonia (1997). An example of time records of the streamwise velocity component, and their derivatives obtained in a field experiment at $Re_\lambda = 10^4$ is shown in Fig. 1.17 in Tsinober (2009). The increasingly intermittent behavior of the signal with the derivative order is seen quite clearly. Also shown are records for the enstrophy ω^2, total strain $2s^2 \equiv 2s_{ij}s_{ij}$ and their surrogate $(\partial u_1/\partial x_1)^2$, and enstrophy production $\omega_i\omega_j s_{ij}$, $s_{ij}s_{jk}s_{ki}$ and their surrogate $(\partial u_1/\partial x_1)^3$. The experiment was performed in the atmospheric surface layer at a height 10 m in approximately neutral (slightly unstable) conditions.

A qualitative summary is that small scale intermittency of turbulence is associated with its spotty (spatio-temporal) *structure* which among other things is manifested as a *particular* kind of non-Gaussian behavior of turbulent flows. This deviation from Gaussianity, increases with both (1) increasing the Reynolds number and

[2]However, note that dissipation is highest in regions where strain—not vorticity—is high. Likewise the enstrophy production, $\omega_i\omega_i s_{ij}$, is small where strain production, $s_{ij}s_{jk}s_{ki}$, is large and versa, see Fig. 7.2 above.

(2) decreasing the 'scale'. In other words, intermittency involves two (not independent) aspects of turbulent flows—their structure/geometry and statistics. These two aspects are reflected in attempts to 'define' intermittency, see references in Tsinober (2009).

9.1.3 Measures/Manifestations of Intermittency

There are numerous quasi-Gaussian manifestations of turbulent flows, for some see Sect. 7.2.2 in Chap. 7, Tsinober (2009). Hence the importance of parameters related to the non-Gaussian nature of turbulence in various contexts of intermittency and structure(s). It is important that intermittency implies non-Gaussianity, but not necessarily vice versa—practically any parameter can at most indicate the degree of intermittency of a flow *already* known to be intermittent. A statistical measure such as flatness or some similar intermittency factors may deviate strongly from a Gaussian value without any intermittency in the flow field. The simplest example is the Gaussian field itself, which by definition lacks any intermittency. However, any nonlinear function (or functional) of a variable, which is Gaussian, is non-Gaussian. For instance enstrophy, dissipation, pressure, etc. of a Gaussian velocity field possess *exponential* tails and their flatness is quite different from 3. For example, for a Gaussian velocity field $F_G(\omega^2) = \langle\omega^4\rangle/\langle\omega^2\rangle^2 = 5/3$ and $F_G(s^2) = \langle s^4\rangle/\langle s^2\rangle^2 = 7/5$. But this by no means indicates that, for a Gaussian velocity field, these quantities are intermittent as claimed sometimes. Moreover, the flatness of enstrophy is *larger* than that of total strain, $F_{\omega^2} - F_{s^2} = 4/15$. Similarly, the Reynolds stress $u_i u_j$ exhibits 'intermittency'. The main contribution to this intermittency comes from the fact that $u_i u_j$ is a product of two random variables both distributed close to Gaussian. For example, the PDF of the $u_1 u_2$ of the strongly intermittent signal obtained by Lu and Willmarth (1973) in a turbulent boundary layer is strongly non-Gaussian. However, the PDF of $u_1 u_2$ is approximated with high precision by assuming both u_1 and u_2 to be Gaussian with a correlation coefficient between them adjusted from the experiment -0.44.[3]

Passive objects (scalars like heat, vectors like magnetic field) in a random velocity field (real or artificially prescribed) are nonlinear functionals. of the velocity field and forcing/excitation. Therefore, even when both the velocity field and forcing are Gaussian the field of a passive object is expected to be strongly non-Gaussian as usually (but not always) is the case (Majda and Kramer 1999). Such *kinematic* intermittency is observed in a great number of theoretical and some experimental works. The term 'kinematic' is used here in the sense that there is no relation to the dynamics of fluid motion, which does not enter in the problems in question, and the velocity field is prescribed and often assumed to be Gaussian.

[3]The above examples may serve as a warning that multiplicative models enable to produce intermittency for a purely nonintermittent field as is the Gaussian velocity field. See Zeldovich et al. (1990) on interesting observations on this and related matters.

Odd Moments Any odd moment of a Gaussian variable vanishes, for example skewness $S_G(a) \equiv \langle a^3 \rangle / \langle a^2 \rangle^{3/2} = 0$. Therefore, odd moments are very sensitive to deviations from Gaussianity, so that non-zero odd moments may be especially good indicators of intermittency. Build-up of odd moments is a result of both the (kinematic) evolution of a passive field in any random velocity field and the *dynamics* of turbulence itself. In the latter case, non-vanishing odd moments are the most important, dynamically significant manifestations of non-Gaussianity, i.e. they reflect *directly* the dynamic aspects of intermittency. The most prominent odd moments are the third order structure function for longitudinal velocity increments $S_3^{\|} = \langle \{[\mathbf{u}(\mathbf{x}+\mathbf{r}) - \mathbf{u}(\mathbf{x})] \cdot (\mathbf{r}/r)\}^3 \rangle$ entering the 4/5 law, the enstrophy production $\langle \omega_i \omega_k s_{ik} \rangle$ and the third order moment of the strain tensor $\langle s_{ij} s_{jk} s_{ki} \rangle$. Note that all these and other odd moments vanish in a Gaussian velocity field. In contrast similar "odd" moments for passive objects, $\langle G_i G_k s_{ik} \rangle$, $\langle B_i B_k s_{ik} \rangle$ do not vanish. Hence, the passive objects in some sense are "more intermittent".

We remind that the non-Gaussian nature of genuine turbulent flows and of passive objects is qualitatively different just like is intermittency in a great variety of physically different systems.

Scaling Exponents and PDFs It is commonly believed that among the manifestations of the small scale intermittency in the commonly defined inertial range[4] is the experimentally observed deviation of the scaling exponents for structure functions $S_p^{\|} = \langle \{[\mathbf{u}(\mathbf{x}+\mathbf{r}) - \mathbf{u}(\mathbf{x})] \cdot (r/r)\}^p \rangle$ for $p > 3$ from the values implied by the Kolmogorov theory, i.e. anomalous scaling, which in turn is due to rare strong events. Namely, $S_p^{\|}(r) \propto r^{\zeta_p^{\|}}$, where $\zeta_p^{\|} = p/3 - \mu_p < p/3$ is a convex nonlinear function of p.

However, there are major problems with scaling as follows.

First, there exists no one-to-one relation between simple statistical manifestations and the underlying structure(s) of turbulence. Moreover, qualitatively different phenomena can and do possess the same set of scaling exponents, so that one needs more subtle statistical characterizations of turbulence structure(s) and intermittency. For example, until recently one of the common beliefs was that the observed vortex filaments/worms are mainly responsible for the phenomenon of intermittency understood as anomalous scaling. However, it appears that this is not the case, see the evidence and references given in Tsinober (1998a, 2009). More specifically, the problem with the intermittency in the conventionally defined inertial range (CDIR) and the related anomalous scaling is in the object (CDIR) itself as it is ill defined and does not exist in reality. Hence the problem with theories attempting

[4]A last example is the paper by Dowker and Ohkitani (2012), see also references therein in which the anomalous scaling is identified with intermittently as is done multitude of papers.

One more example is in Seiwert et al. (2008) who studied *the decrease of intermittency in decaying rotating turbulence* via looking at the scaling of the longitudinal velocity structure functions, up to order $q = 8$. This decrease can be explained by suppression of strong dissipative events in the presence of rotation.

explanations of intermittency in this nonexistent phenomenon, e.g. breakdown coefficients/multipliers (Novikov 1974, 1990a), multi-fractals (Frisch 1995) and the so-called 'hierarchical symmetry' (She and Zhang 2009).

There are no 'corrections' to the scaling exponent in $4/5$ law—it is an exact consequence of NSE. However, as (i) it manifests the non-Gaussian nature of turbulence and (ii) the PDFs of the longitudinal velocity increments especially at small r have flaring tails, i.e. hanging far above the Gaussian PDF, the $4/5$ law should be considered as related to intermittency. This shows that 'intermittency corrections' are not that reliable as indicators of intermittency, if at all. Reminding that recent experiments at high Reynolds numbers showed that the $4/5$ law is not a pure inertial relation (which is one of the manifestations of the ill posedness of the CDIR) since the PDFs of the velocity increments contain strong dissipative events with nonnegligible contributions to the structure functions $S_p^{\|}(r)$ increasing with the order and among them $S_3^{\|}(r)$.

The next example is represented by numerous models attempted to reproduce the anomalous scaling. A partial list of references is given in Sreenivasan and Antonia (1997) and Tsinober (1998b). These models followed the Kolmogorov (1962) refined similarity hypothesis (RSH) in which the mean dissipation $\langle\epsilon\rangle$ was replaced by 'local' dissipation ϵ_r averaged over a region of size r. The scaling exponents obtained in all of these models are in good agreement with the experimental and numerical evidence, e.g. these models exhibit the same scaling properties (and some other such as PDFs) as in real turbulence. It is noteworthy that many of these models are based on qualitatively different premises/assumptions and with few exceptions have no direct bearing on the Navier–Stokes equations. **Therefore the success of such models can hardly be evaluated on the basis of how well they agree with experiments. Phenomenology and models only will hardly be useful and convincing, since almost any dimensionally correct model, both right or wrong, will lead to correct scaling without appealing to NSE and/or elaborate physics.** For example, there exist many theories which produce the $k^{-5/3}$ energy spectrum for qualitatively and/or physically different reasons. A recent example is a *suggestion that the spectrum of fully developed turbulence is determined by the equilibrium statistics of the Euler equations and that a full description of turbulence requires only a perturbation, small in some appropriate metric, of a Gibbsian equilibrium* (Chorin 1996). The most common justification for the preoccupation with such models is that they (at least some of them) share the same basic symmetries, conservation laws and some other general properties, etc. as the Navier–Stokes equations. The general belief is that this—along with the diversity of such systems (there are many having nothing to do with fluid dynamics, e.g. granular systems, financial markets, brain activity)—is the reason for the above mentioned agreement. However, this is not really the case, e.g. in Kraichnan (1974) a counter example of a *'dynamical equation is exhibited which has the same essential invariances, symmetries, dimensionality and equilibrium statistical ensembles as the Navier–Stokes equations but which has radically different inertial-range behavior'*! The majority of models exhibit temporal chaos only. Therefore, such and most other models hardly can be associated with the intermittency of real fluid turbulence, which involves essentially spatial chaos as well. Again, for the above reasons the agreement between such models and experiments (both laboratory and numerical) cannot be used for evaluation of the success of such models. There are proposals to use two sets of independent exponents $\zeta_p^{\parallel}$ and $\zeta_p^{\perp}$ (Chen et al. 2003) and there exist other 'universality' proposals involving 'much more' scaling exponents, see e.g. Biferale and Procaccia (2005), Frisch (1995) and references therein.

Scaling laws alone are not necessarily theories. With all the attractiveness of scaling, turbulence phenomena are infinitely richer than their manifestation in scaling and related things. Most of these manifestations are beyond the reach of phenomenology. Phenomenology is inherently unable to handle the structure of turbulence in general, and phase and geometrical relations in particular, to say nothing of dynamical features such as build up of *odd* moments, interaction of vorticity and strain resulting in positive net strain and enstrophy production/predominant vortex stretching. It seems that there is little promise for progress in understanding the basic physics of turbulence in going on dealing with scaling and related matters only, without looking into the structure and, where possible, basic mechanisms which are *specific* to turbulent flows. In fact, the main question of principle which should have been asked long ago is: Why on earth should we perform so many elaborate measurements of various scaling exponents without looking into the possible concomitant physics and/or without asking why and how more precise knowledge of such exponents, even assuming their existence, can aid our understanding of turbulent flows? This is not to say that one has to abandon the issues of scaling. An example of affirmative answer is given in Sect. 7.1 concerning the ill-posedness of the concept of inertial range.

Second, as discussed in the previous section the very existence of scaling exponents in statistical sense (as, e.g. for various structure functions or corresponding PDFs, etc.) which is taken for granted, is a problem by itself.

A similar question arises in respect with multifractality which was designed to 'explain' the 'anomalous' scaling, since there is no direct experimental evidence on the multifractal structure of turbulent flows. So there is a possibility that multifractality in turbulence is an artifact (see Frisch 1995, p. 190). Moreover, in reality multifractality in fact is a kind of description of finite Reynolds number effect at whatever large *Re* due to ill posedness of the inertial range as mentioned several times above.

The PDFs of an intermittent variable are quite useful as, e.g. they carry the information showing that extremely small and extremely large values are much more likely than for a Gaussian variable. However, they contain no information on the structure of the underlying weak and strong events, nor on the structure of the background field. Hence, the same PDFs can have qualitatively different underlying structure(s) of the flow, i.e. 'how the flow looks'. Similar PDFs of some quantities can correspond to qualitatively different structure(s) and quantitatively different values of Reynolds number, see references in Tsinober (1998a, 1998b, 2009). For example, the qualitative difference in the behavior and properties of regions dominated by strain and those with large enstrophy cannot be captured by such means and other conventional measures of intermittency. Also the PDFs, like scaling exponents, do not allow us to infer much about the underlying dynamics. This, however, is true of 'conventional' PDFs like those of velocity increments, but not of any PDFs such as those directly associated with geometrical flow properties.

Note that the largest deviation from Gaussianity occurs at small scales. In this sense, the field of velocity derivatives, $\partial u_i/\partial x_k$, is more intermittent than the field of velocity, u_i, itself. One of the possible reasons for this is in the different nature of nonlinearity at the level of velocity field, i.e. in the Navier–Stokes equations and, for example, in the equation for vorticity.

9.1.4 On Possible Origins of Small Scale Intermittency

At the present state of matters the issue is pretty speculative, and an example of 'ephemeral' collection of such is given below.

As one of manifestations of turbulence structure(s), intermittency has its origins in the structure of turbulence, see next section. Therefore we briefly address here the issue on possible origins of intermittency. There are roughly two kinds of origins of intermittency: kinematic and dynamic.

Before proceeding we reiterate again that non-Gaussianity and intermittency are not synonymous just like the origins of non-Gaussian statistics in various systems and genuine turbulence are generally quite different even qualitatively. Therefore, it is misleading to 'explain' such properties of genuine turbulence by analogy with non-Gaussian behavior of, e.g. Burgers and/or restricted Euler equations. An important point is that these are integrable equations, and exhibit random behavior only under random forcing and or initial conditions, otherwise their solutions are not random and should be distinguished from problems involving genuine turbulence. Navier–Stokes equations at sufficiently large Reynolds number have the property of intrinsic mechanisms of becoming complex without any external aid including strain and vorticity production. There is no guarantee that the outcome, e.g. such as structure(s) is the same from, e.g. natural "self-randomization" and with random forcing and even with different kinds of forcing. Moreover there is evidence that the outcome is indeed different.

Direct and Bidirectional Interaction/Coupling Between Large and Small Scales
As discussed, direct and bidirectional interaction/coupling between large and small scales is one of the elements of the nonlocality of turbulence. It is both of kinematic and dynamic nature.

The first recognized manifestation of such interaction is that the small scales do not forget the anisotropy of the large ones. There is a variety of mechanisms producing and influencing the large scales: various external constraints like boundaries with different boundary conditions, including the periodic ones, initial conditions, forcing (as in DNS), mean shear/strain, centrifugal forces (rotation), buoyancy, magnetic field, external intermittency in partially turbulent flows, etc. Most of these factors usually act as organizing elements, favoring the formation of coherent structures of different kinds (quasi-two-dimensional, helical, hairpins, etc.). These, as a rule, large scale features depend on the particularities of a given flow that are not universal. Therefore the direct interaction between large and small scales leads to 'contamination' of small scales by the large ones, e.g. the edges of large scale structures are believed to be responsible for such 'contamination' in any kind of flow. This contamination is unavoidable even in homogeneous and isotropic turbulence, since there are many ways to produce such a flow, i.e. many ways to produce the large scales. It is the difference in the mechanisms of large scale production which 'contaminates' the small scales. Hence, non-universality.

The direct and bidirectional interaction/coupling of large and small scales, i.e. nonlocality, is a generic property of all turbulent flows and one of the main reasons for small scale intermittency, non universality, and quite modest manifestations of scaling. This dates back to the famous Landau remark stating that the *important part will be played by the manner of variation of ε over times of the order of the periods of large eddies (of size ℓ)* (Landau and Lifshits 1944, see 1987, p. 140).

"Near" Singularities It is not known for sure whether Navier–Stokes equations at large Reynolds numbers develop a genuine singularity in finite time, though there no evidence in favor of this, so the term "near" singularities is just another term for strong events not necessarily just dissipative. Nevertheless, it seems a reasonable speculation that the 'near' singularities *trigger topological change and large dissipation events; their presence is felt at the dissipation scales and is perhaps the source of small scale intermittency* (Constantin 1996).

In any case, the 'near' singular objects *may* be among the origins of intermittency of a dynamical nature.[5] However, there is a problem with two-dimensional 'turbulence'. Namely, in this case everything is beautifully regular, but there is intermittency in the sense of the above definitions, with the exception of scaling exponents for velocity structure functions and corresponding quasi-Gaussian behavior. However, non-Gaussianity is strong at the level of velocity derivatives of a second order. Hence the possible formation of singularities in 3D is not necessarily the underlying reason for intermittency in 3D turbulence. Another example relates to modified Navier–Stokes equations such as those using hyperviscosity replacing the Laplacian by a higher order operator $(-1)^{h+1}\nabla^{2h}$ with $h > 1$ with the underlying assumption that this manipulation changes only the small scales. In this case too everything is beautifully regular too for $h > 5/4$, i.e. the solution remains regular for all times and any Reynolds number (Ladyzhenskaya 1975; Lions 1969) and *some* features of turbulence are reproduced well (such $k^{-5/3}$ spectrum) including intermittency, but its structure(s) appear quite different from those as for true NSE.

Multiplicative Noise, Intermittency of Passive Objects in Random Media It has been known for about thirty years that passive objects (scalars, vectors) exhibit

[5] We mean singularities which appear at random in space and time and not in a strictly periodic (and fully coherent and mutually amplifying) fashion as in DNS with periodic boundary conditions.

'anomalous scaling' behavior and other strong manifestations of intermittency even in pure Gaussian random velocity field, see references in Tsinober (2009). These are dynamically linear systems, but they are of the kind which involve the so-called multiplicative 'noise', i.e. the coefficients in the equations that depend on the velocity field. Therefore, statistically they are 'nonlinear', since the field of passive objects is a nonlinear functional of the velocity field. Therefore, passive objects exhibit strong deviations from Gaussianity. In such systems, intermittency results either from external pumping (forcing term on RHS of the equations), or in systems without external forcing from instability (self-excitation) of a passive object in a random velocity field under certain conditions.

The velocity field does not 'know' about the passive objects. In this sense, problems involving passive objects are kinematic in respect with the velocity field in real fluid turbulence. They may reflect the contribution of kinematic nature in real turbulent flows. In view of the recent progress in this field it was claimed that *investigation of the statistics of the passive scalar field advected by random flow is interesting for the insight it offers into the origin of intermittency and anomalous scaling of turbulent fluctuations* (Pumir et al. 1997), for later references see Tsinober (2009). More precisely it offers an insight into the origin of intermittency and anomalous scaling of fluctuations in random media generally and independently of the nature of the random motion (Zeldovich et al. 1990), i.e. it gives some insight into the contributions of kinematic nature, but does not offer much regarding the specific dynamical aspects of strong turbulence in fluids. Moreover, anomalous diffusion and scaling of passive objects occurs in purely laminar flows in Eulerian sense (E-laminar flows) as a result of Lagrangian chaos (L-turbulent flows), i.e. intermittency of passive objects may even have nothing to do with the random nature of fluid motion in Eulerian setting.

Thus in real turbulent flows there are two contributions to the behavior of passive objects, kinematic and dynamic. It seems hopeless to separate them in any sense.

Summarizing, intermittency specifically in genuine fluid turbulence is associated mostly with some aspects of its spatiotemporal structure, especially the spatial one. Hence, the close relation between the origin(s) and meaning of intermittency and structure of turbulence. Just like there is no general agreement on the origin and meaning of the former, there is no consensus regarding what are the origin(s) and what turbulence structure(s) really mean. What is definite is that turbulent flows have lots of structure(s). The term structure(s) is used here deliberately in order to emphasize the duality (or even multiplicity) of the meaning of the underlying problem. The first is about how turbulence 'looks'. The second implies the existence of some entities. Objective treatment of both requires use of some statistical methods. It is thought that these methods alone may be insufficient to cope with the problem, but so far no satisfactory solution was found. One (but not the only) reason—as mentioned—is that it is not so clear what one is looking for: the objects seem to be still elusive. For example, there is still a non-negligible set of people in the community that are in a great doubt that the concept of coherent structure is much different from the Emperors's new Clothes.

9.2 What Is(Are) Structure(s) of Turbulent Flows? What We See Is Real. The Problem Is Interpretation

What we see is real. The problem is interpretation as there is even a problem of defining of "seeing".

The issue is pretty speculative, and an example of 'ephemeral' collection of such is given below. We have to admit at this stage that structure(s) is(are) just an inherent property of turbulence. Structureless turbulence is meaningless.

The difficulties of definition what the structure(s) of turbulence are of the same nature as the question about what is turbulence itself. So before and in order to 'see' or 'measure' the structure(s) of turbulence one encounters the most difficult questions such as: what is (say, dynamically relevant) structure?, Structure of what? Which quantities possess structure in turbulence? What is the relation between structure(s) and 'scales'—unfortunately both ill defined? Can structure exist in 'structureless' (artificial) pure random Gaussian fields? Which ones? All this—like many other issues—are intimately related to the skill/art to ask the right and correctly posed questions. These impossibly difficult questions are made not easier due to quite a bit of turbulence in terminological aspects and terminological abuse by use of a variety of ill defined terms (eddies, worms, sheets, tubes, pancakes, ribbons, vortex or vortical structures/filaments, vortons, 'eigensolutions', significant shear layers, etc. Some people would include in this list also coherent structures frequently used as synonyms of components of some decompositions or similar "executions" of the real flow field, which are usually followed by studies—sometimes pretty sophisticated—of their interaction not necessarily reflecting any physics, at least as concerns physical space. One of the popular games of this kind is looking for confirmation of *the classical energy cascade picture,* such as in the latest examples in Aluie (2012), Leung et al. (2012) and references therein.

The main starting point here is at the end of the previous section: just like there is no general agreement on the origin and meaning of intermittency, there is no consensus regarding what are the origin(s) and what turbulence structure(s) really mean. What is definite is that turbulent flows have lots of structure(s). The term structure(s) is used here deliberately in order to emphasize the duality (or even multiplicity) of the meaning of the underlying problem. The first is about how turbulence 'looks'. The second implies the existence of some entities. This follows by more serious issues which unfortunately are mostly even not properly posed.

The meaning of structure(s) depends largely on what is meant by turbulence itself, and especially structure(s) of the particular field one is looking. For example, the velocity field may have no structure, but the passive tracer may well have a pretty nontrivial one, simple laminar Eulerian velocity field (E-laminar) creates complicated Lagrangian field (L-turbulent). Purely Gaussian, i.e. 'structureless' velocity field creates structure in the field of passive objects. The structure(s) seen in the velocity field depend on the motion of the observer. Finally, what is called "coherent structures" or "organized motion", which has been rediscovered many times, may be not directly related to the turbulent nature of the flow (such as mixing layer), but are rather a result of large scale instability of the flow as 'whole' (zooming out).

9.2.1 On the Origins of Structure(s) of/in Turbulence

This question—in some sense—is a 'philosophical' one. But its importance is in direct relation to even more important questions about the origin of turbulence itself. Hence again an 'ephemeral' collection of such possible reasons/causes of structure(s) in turbulence flows keeping in mind that structure(s) is (are) just an inherent property of turbulence. There is no turbulence without structure(s).

Instability As mentioned in Chap. 2, the most commonly accepted view on the origin of turbulence is flow instability. An additional factor is that instability is considered as one of the origins of structure(s) in/of turbulence. However, this latter view requires to assume that turbulence has a pretty long 'memory' of or, alternatively, that the 'purely' turbulent flow regime, i.e. at large enough Reynolds numbers, has instability mechanisms similar to those existing in the process of transition from laminar to turbulent flow state. The problem is that speaking about (in)stability requires one to define the state of flow (in)stability of which is considered, which is not a simple matter in the case of a turbulent flow.

Note the observation made by Goldshtik and Shtern (1981): *The fact that the phenomenon of intermittency and structures are observed in the proximity of the outer boundary of turbulent flow or in close to the wall and in the small scale "tail" of turbulent flow flows, i.e. when the characteristical Reynolds numbers are relatively not large, prompts an assumption that "structureness" is associated with mechanisms of turbulence origins.* This may be an underlying reason of some similarity between some flow patterns ("structures") in transitional and developed turbulent flows. This idea appeared in a number of subsequent publications, e.g. Blackwelder (1983), Pullin et al. (2013) and references therein. It should be stressed that even if the above hypothesis is true there in no escape from nonlocal effects!

Emergence Another less known view holds that structure(s) emerge in large Reynolds number turbulence out of 'purely random structureless' background, e.g. via the so-called inverse cascades or negative eddy viscosity. Among the spectacular examples, are the 'geophysical vortices' in the atmosphere, and ocean, as well as astrophysical objects. Another example is the emergence of coherent entities, such as vortex filaments/worms and other structure(s), out of an initially random Gaussian velocity field via the NSE dynamics, for examples see references in Tsinober (2009).

Anderson (1972) emphasizes *the concept of 'broken symmetry', the ability of a large collection of simple objects to abandon its own symmetry as well as the symmetries of the forces governing it and to exhibit the 'emergent property' of a new symmetry*. One of the difficulties in turbulence research is that no objects simple enough have been found so far such that a collection of these objects would adequately represent turbulent flows. It is not clear how meaningful is the very question on the existence of such objects.

It 'Just Exists', or Do Flows Become or Are They Are 'Just' Possessing Structure? *To the flows observed in the long run after the influence of the initial conditions has died down there correspond certain solutions of the Navier–Stokes equations. These solutions constitute a certain manifold $\mathcal{M} = \mathcal{M}(\mu)$ (or $\mathcal{M} = \mathcal{M}(Re)$) in phase space invariant under phase flow* (Hopf 1948). *Kolmogorov's scenario was based on the complexity of the dynamics along the attractor rather than its stability* (Arnold 1991), see also Keefe (1990a), Keefe et al. (1992).

This view is a reflection of one of the modern beliefs that the structure(s) of turbulence—as we observe in *physical space*—is (are) the manifestation of the generic structural properties of mathematical objects in *phase space*, which are called attractors and which are invariant in some sense. In other words, here the structure(s) are assumed to be 'built in' the turbulence independently of its origin, hence the tendency to universality.[6] It is noteworthy that the assumed attractor existence makes sense for statistically stationary turbulent flows. However, for flows which are not such, e.g. decaying turbulent flows past a grid or a DNS simulated flow in box the attractor is trivial. Nevertheless, these flows possess many properties which are essentially the same as their statistically stationary counterparts provided that their Reynolds numbers are not too small ($Re_\lambda \geq 10^2$).

The above refers to the dynamical aspects of real turbulent flows. We mention again here also the

Emergence of Structures in Passive Objects in Random Media In which the velocity field and the external forcing are prescribed. Whatever their nature—even Gaussian—structure is emerging in the field of passive objects (Zeldovich et al. 1990; Ott 1999 and references therein).

9.2.2 How Does the Structure of Turbulence 'Look'?

For long time the first and the only impression/answer to this question was obtained by employing visualization techniques. First in experiments using mainly passive markers and later using the DNS simulations looking mainly at objects bounded by isosurfaces of some quantity, such as enstrophy, ω^2.

The first important result was that even turbulence which is 'homogeneous' and 'isotropic' has structure(s), i.e. contains a variety of strongly localized events. The primary evidence is related to spatial localization of subregions with large enstrophy, i.e. intense vorticity, which are organized in long, thin tubes-filaments-worms. Some evidence was obtained about regions with large strain, $s_{ij}s_{ij}$, i.e. dissipation, being sheet-like objects with very sharp edges (razors/flakes), see references in Tsinober (2009) for both.

[6] In the strange attractor theory, the experimental measurements are viewed as projections of these attractors onto low dimension that correspond to these measurements.

The relatively simple appearance of the observed 'structures' is due simple techniques as looking at isosurfaces of some quantity, e.g. enstrophy ω^2, with thresholding using conditional sampling techniques. This is how the first evidence of concentrated vorticity/filaments/worms was obtained.

This prompted a rather popular view that turbulence structure(s) is (are) simple in some sense and that essential aspects of turbulence structure and its dynamics may be adequately represented by a random distribution of simple (weakly interacting) objects.

In particular, it is commonly believed that most of the structure of turbulence is associated with and is due to various strongly localized intense events/structures, e.g. mostly regions of concentrated vorticity so that *turbulent flow is dominated by vortex tubes of small cross-section and bounded eccentricity* (Chorin 1994, p. 95), for other quotes and references see Tsinober (1998a, 2009), and that these events are mainly responsible for the phenomenon of intermittency. It is demonstrated in Tsinober (1998a, 2009) that such views are inadequate. It appears that—though important—these structures are not the most dynamically important ones and are the consequence of the dynamics of turbulence rather than its dominating factor. Namely, regions other than those involving concentrated vorticity such as: (i) 'structureless' background, (ii) regions of strong vorticity/strain (self) interaction and largest enstrophy and strain production dominated by large strain rather than large enstrophy, and (iii) regions with negative enstrophy production are all dynamically significant and in some important respects more significant than those with concentrated vorticity, strongly non-Gaussian, and possess structure. Due to the strong nonlocality of turbulence in physical space all the regions are in continuous interaction and are strongly coupled. A similar statement can be made regarding the so called streamwise vortices observed in many turbulent flow. Moreover, as described in Chap. 7 the anomalous scaling is due to strong dissipative events, i.e. large strain s^2, so that the conventionally defined inertial range is an ill defined concept and turning it out of existence.

To emphasize, the above conclusions are the outcome of the use of quantitative manifestations of turbulence structure, which just like intermittency are in the first place of statistical nature independently of how specifically the individual structures look and whether they do exist at all.

Though the isosurfacing/thresholding approach is useful and 'easy', it is inherently limited and reflects at best the simplest aspects of the problem. Even for characterization of *some* aspects of the *local* (i.e. in a sense 'point'-wise) structure of the flow field in the frame following a fluid particle requires *at least two* parameters.[7] Therefore attempts to *adequately* characterize finite scale structure(s) by one parameter only are unlikely to be successful. The one parameter approaches are not

[7]These are $Q = 1/4(\omega^2 - 2s^2)$ and $R = -1/3(s_{ij}s_{jk}s_{ki} + 3/4\omega_i\omega_j s_{ij})$. Here Q is the second—and R is the third invariant of the velocity gradient tensor $\partial u_i/\partial x_j$. The first is vanishing due to incompressibility.

A similar problem arises when attempting to characterize structure(s) of turbulent flows using two-point information but based on a single velocity component only, e.g. longitudinal structure

made any better (but rather more misleading) by adding to thresholding and isosurfacing some decomposition, e.g. Leung et al. (2012) and references therein. One can study some geometrical issues of the isosurfaces in the filtered fields and even the "interaction" between such "structures" belonging to the fields corresponding to different filter bands with the remaining acute problem as to how all this is related to the whole flow field in the real physical space. More generally, the problem is related to pattern recognition and requires defining a conditional sampling scheme involving more than two parameters. This scheme is in turn based on what a particular investigator thinks are the most important physical processes, features, etc. This in turn opens a Pandora box of possibilities and contains an inherent element of subjectivity and arbitrariness, since the physics of turbulence is not well understood. In this sense, the circle is closed: in order to objectively define and educe some structure, one needs clear understanding of the physics of turbulence, which, it is in turn believed, can be achieved via study of turbulence structure(s).

There are other serious problems in observations of individual structures obtained via isosurfacing and thresholding or similar and alike.

First, the "boundaries" of flow regions isolated by such methods cannot be qualified as "natural" in any sense and serve as a technical means only. One cannot take such an approach for granted as reliable for getting the "natural" boundaries of these structures and serving simultaneously as definitions of those "structures". Both can be qualified as wishful thinking at best. On the other hand, there exists a number of attempts to define what is, e.g. a coherent structure, a vortex, etc., see references in Malm et al. (2012) and also Monin (1991), Townsend (1987), Bonnet (1996) and Holmes et al. (1996).

Second, these structures are just single time snapshots in space having no identity beyond the particular time moments of their (infinitesimal) life time, so that one cannot observe their time evolution. The latter difficulty is nontrivial, because one ventures to deal with *finite* objects. Namely, even having defined such a finite relatively simple object at some time moment one is loosing it in a pretty short time even if there is a possibility to follow this object as in case of purely Lagrangian objects. This is illustrated in Fig. 9.2. It is for these reasons people produce statistics out of collections of "similar objects" obtained from snapshots at the same and different time moments via isosurfacing and thresholding and other tricks to justify these "surgeries", etc. But the painful question is how really "similar" are all these if they are typically defined by one parameter only? It is almost obvious that such kind of "statistical" processing is killing most of essential features of the real "structure" and leaves the question of relevance, say the dynamical one, of these "structures" at best open.

The above points to acute problems in defining instantaneous structures not to mention studying them, though it is commonly assumed that there exist instantaneous "structures" which are in some sense "key" objects from some point of view

---

functions $S_n^{\|}(r)$, since such an approach does not 'know' (almost) anything about the two other velocity components.

Likewise there are many attempts to handle 'structures' even with just steamwise velocity component time series.

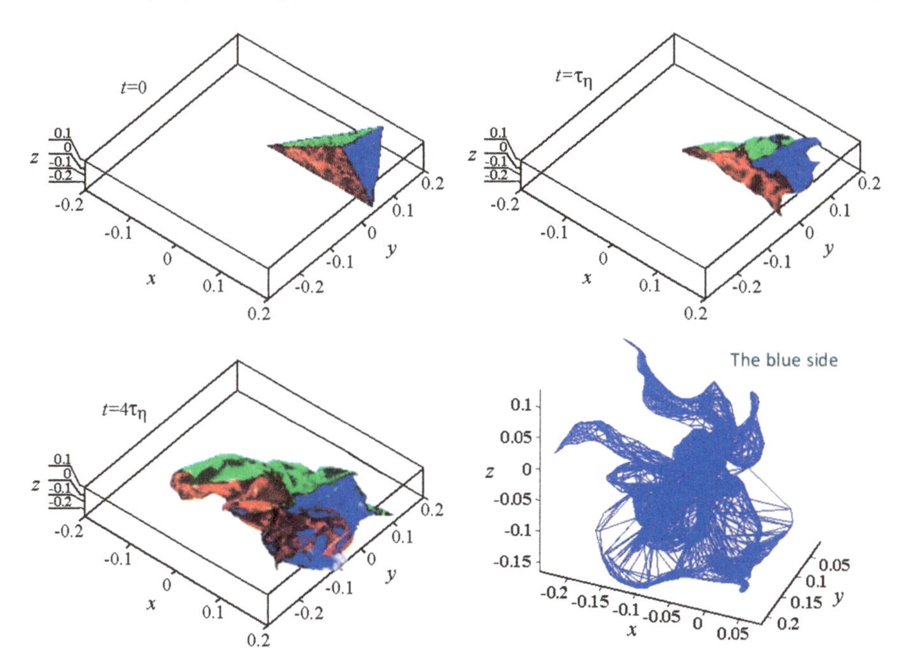

Fig. 9.2 Evolution of a tetrahedron with edge of $\approx 4\eta$ at $t = 0$ using the data base of Johns Hopkins University (Li et al. 2008), for a 1024^4 space-time history of a direct numerical simulation of isotropic turbulent flow in incompressible fluid in 3D. Courtesy Beat Luethi. It seen that a simple Lagrangian tetrahedron, i.e. consisting of fluid particles, becomes not so simple after just one Kolmogorov time scale τ_η and turns into a non-trivial object in time of few Kolmogorov time scales

and that these unknown objects even govern the dynamics of the flow. Claims of this kind are pretty frequent, but without much—if any—explanation/justification or whatever.

As concerns the individual structure it was already mentioned that the regions of concentrated vorticity are of limited dynamical relevance. First, these regions are characterized by approximate balance between enstrophy production $\omega_i\omega_k s_{ik}$ and its viscous destruction in a way similar to that of Tennekes and Lumley balance, see Tsinober (2009). Second, they belong to the category of flow patterns with predominant alignment of vorticity and the strain eigenvector $\boldsymbol{\lambda}_2$ corresponding to the intermediate strain eigenvalue, Λ_2. However, the major contribution to the enstrophy production comes from the regions with the $\boldsymbol{\omega}, \boldsymbol{\lambda}_1$ alignment (corresponding to the largest strain eigenvalue, Λ_1) and in which there is no approximate balance between enstrophy production $\omega_i\omega_k s_{ik}$ and its viscous destruction with strong dominance of $\omega_i\omega_k s_{ik}$. Moreover, the regions with the $\boldsymbol{\omega}, \boldsymbol{\lambda}_1$ alignment comprise a large part of those where the vorticity/strain interaction is strongest, see Chap. 7 above and Chap. 6 in Tsinober (2009).

Recently there is some trend of reviving and ascribing some excessive importance to thin shear layers (Hunt et al. 2010; Elsinga and Marusic 2010; Worth and Nickels 2011; Ishihara et al. 2011 and references therein). However, they are not more relevant than the "worms" as they belong to the same category with predominant alignment of vorticity and the strain eigenvector corresponding to Λ_2 the intermediate strain eigenvalue Λ_2,[8] whereas as mentioned the most dynamically active are flow patterns with predominant $\boldsymbol{\omega}, \boldsymbol{\lambda}_1$ alignment, e.g. as concerns enstrophy production and other essential nonlinear processes, see Chap. 7 above (Tsinober 2009 and references therein).

A final note is that though the patterns with predominant alignment of vorticity and the strain eigenvector corresponding to the intermediate strain eigenvalue, Λ_2 (worms, shear layers and more involved patterns) are statistically dominant they are not the most dynamically relevant. In other words, statistical dominance is not synonymous to dynamical relevance. The qualification of "shear layer" as (several previous "key" objects, e.g. worms) belongs to the category of oversimplified concepts and vague terminology—as mentioned more preferable are well defined strain and vorticity. The oversimplification (thin layers!) is seen clearly from the above as neglecting important issues of geometrical nature among others, e.g. the aspects of alignments of vorticity and the eigenframe of the rate of strain tensor.

9.2.3 Structure Versus Statistics

The 'not objective enough' nature of a variety of conditional sampling procedures resulted in a whole 'zoo' of 'structures' in different turbulent flows, which some people believe to be significant in some sense, but many do not. The zoo seems to have a tendency to grow at least exponentially with the introduction of multiscale approaches, but one cannot help reminding the question by Kadanoff (1986) when the mulifractal "formalism" was just born: *Where is the physics?* Among the reasons for such skepticism is some evidence that the attempts at adequate representation of such a complicated phenomenon like turbulence as a collection of simple objects/structures only are unlikely to succeed. As mentioned, until recently it was believed that concentrated vorticity/filaments is the dominating structure in turbulent flows in the sense that most of the structure of turbulence is associated with and is due to regions of concentrated vorticity. It appears that—though important—these structures are not the most dynamically important ones and are the consequence of the dynamics of turbulence rather than being its dominating factor. A similar statement is true in respect of recently revived sheer layer.

Nevertheless, some 'objectiveness' can be achieved using quantities appearing in the NSE and/or the equations which are exact consequences of NSE.

The question about what structure(s) of turbulence mean(s) can be answered via a statement of impotence: speaking about 'structure(s)' in turbulence the implication

[8] Indeed, e.g. Ishihara et al. (2011) state on observations of *thin shear layers consisting of a cluster of* ***strong vortex tubes*** *with typical diameter of order* 10η, *where* η *is the Kolmogorov length scale.*

is that there exist something 'structureless', e.g. Gaussian random field as a representative of full/complete disorder. Gaussian field is appropriate/natural to represent the absence of structure in the statistical sense. Hence all non-Gaussian manifestations of turbulent flows can be seen as some statistical signature of turbulence structure(s). This does not imply that an exactly Gaussian field does not necessarily possess any spatial or temporal structures, see, e.g. Fig. 3 in She et al. (1990)—any individual realization of a Gaussian field does have structures. However, an exactly Gaussian field does not possess dynamically relevant structure(s), it is dynamically impotent.

So the next most difficult question is about the relevance/significance of some particular aspect of non-Gaussianity for a specific problem in question. It seems that here one enters the *subjective* realm: the criteria of significance (which is the matter of physics!) are decided by the researchers. However, the following examples show that objective choice of the structure sensitive statistics is dictated by general dynamical aspects of the problem.

For instance, the build up of *odd* moments is an important *specific* manifestation of structure of turbulence along with being the manifestation of its nonlinearity. The two most important examples are the third order velocity structure function $S_3(r) = \langle\{[\mathbf{u}(\mathbf{x}+\mathbf{r}) - \mathbf{u}(\mathbf{x})] \cdot \mathbf{r}/r\}^3\rangle$ and the mean enstrophy production $\langle\omega_i\omega_k s_{ik}\rangle$. The first one is associated with the $-4/5$ Kolmogorov law $S_3(r) = -4/5\langle\epsilon\rangle r$ (Kolmogorov 1941b), which is the first strong indication of the presence of structure in the inertial range showing that both non-Gaussianity and the structure of turbulence are directly related to it's dissipative nature. It is remarkable that the title of this paper by Kolmogorov is *Dissipation of energy in the locally isotropic turbulence.* The $-4/5$ Kolmogorov law clearly overrules the claims that '*Kolmogorov's work on the fine-scale properties ignores any structure which may be present in the flow*' (Frisch 1995, p. 182) and that it is associated with near-Gaussian statistics, see references in Tsinober (2009) among multitude of others. As concerns the near Gaussian statistics it is correct that single point statistics is known to be quite close to the Gaussian one. However, the conclusion that velocity fluctuations are really almost Gaussian is a misconception, not to mention the field of velocity derivatives. This is already seen when one looks at two-point velocity statistics. For instance, in such a case the odd moments are significantly different from zero, e.g. Frenkiel et al. (1979).

The essentially positive value of the mean enstrophy production $\langle\omega_i\omega_k s_{ik}\rangle$, discovered by Taylor (1938a) is the first indication of the presence of structure in the small scales, where turbulence is particularly strongly non-Gaussian and intermittent. The above two examples show that both the essential turbulence dynamics and its structure are associated with those aspects of it's non-Gaussianity exhibited in the build up of odd moments, which among other things means phase and geometrical coherency, i.e. structure. Hence, the importance of odd moments as indicators of intermittency. It is to be noted that the non-Gaussianity found experimentally both in large and small scales is exhibited not only in the nonzero odd moments, but also in strong deviations of even moments from their Gaussian values. Thus both the large and small scales differ essentially from Gaussian indicating that both possess structure.

However, an important point is that probability criteria are insufficient, since *one can find in statistical data irrelevant structures with high probability* (Lumley 1981). In other words the structure(s) should be relevant/significant in some sense. For example, it should be *dynamically* relevant for velocity field, and related quantities such as vorticity and strain. This does not mean that kinematical aspects of turbulence structure(s) are of no importance. For example, *anisotropy* is a typical *kinematic statistical* characteristic of turbulent flows of utmost dynamical significance/impact which hardly can be applied to *individual* structures, e.g. a turbulent flow consisting mostly of 'anisotropic' individual structures can be statistically isotropic. Among the first statistical treatments of turbulence structure is, of course, the first paper by Kolmogorov (1941a), the very title of which is *The local structure of turbulence in incompressible viscous fluid for very large Reynolds numbers.*

The advantage of such an approach is that it allows one to get insights into the *structure* of turbulence without the necessity of knowing much (if anything) about the actual appearance of it's *structures*, since the very question of this kind may well be just meaningless. This is especially important in view of numerous problems/ambiguities in definitions of *individual* structures in turbulent flows, their identification and statistical characterization as well as their incorporation in 'theories'. The main reason is that there exist an intrinsic problem of both defining what the *relevant* structures are, see Bonnet (1996) for references and a review of existing techniques which all are based on statistics anyhow, e.g. of defining extracting/educing and characterizing the so-called coherent structures. For a number of reasons, it is very difficult, if not impossible, to quantify the information on the *instantaneous* structures of turbulent flows into dynamically relevant/significant form. The observed individual structures strongly depend on the method observation/extracting, but more importantly none of them are simple, neither are they weakly interacting between themselves or with the background which in fact is not structureless as assumed by many. Indeed, *you can find structures, essentially arbitrary, which have equal probability to the ones we have latched onto over the years: bursts, streaks, etc... If structures are defined as those objects which can be extracted by conditional sampling criteria, then they are everywhere one looks in turbulence* (Keefe 1990b). For instance, looking at a snapshot of the enstrophy levels of a purely Gaussian velocity field in She et al. (1990) one can see a number of filaments—the irrelevant ones—like those observed in real turbulent flows, i.e. pure Gaussian velocity field has some structure(s) too.

9.2.4 What Kinds of Statistics Are Most Appropriate to Characterize at Least Some Aspects for Turbulence Structure

Returning to isosurfacing and thresholding it should be mentioned that these allow to handle regions (rather than "structures") with some properties of interest, for latest example see Malm et al. (2012). These authors used a measure of vorticity stretching related to the magnitude of the vortex stretching vector $W_i \equiv \omega_{i s_{ij}}$, which is exemplifying that the main criteria should be of dynamical relevance, and sensitiveness to the non-Gaussian properties of turbulence, so that one can speak about statistics weakly sensitive to structure and structure sensitive statistics.

Examples of Statistics Weakly Sensitive to Structure(s) The first examples of this kind are energy spectra in which the phase (and geometric) information is lost. Hence their weak sensitivity to the structure of turbulence. This insensitivity, in particular, is exhibited in the scaling exponents when/if such exist. For example, the famous $-5/3$ exponent can be obtained for a great variety of *qualitatively* different real systems—not necessarily fluid dynamical—and theoretical models, for a partial list of references see Tsinober (2009). One can also construct a set of purely Gaussian velocity fields, i.e. lacking any dynamically relevant structure(s), with any desired length of the $-5/3$ 'inertial' range (Elliott and Majda 1995). An extreme example is a single sharp change in velocity represented in Fourier space has an energy spectrum $\propto k^{-6/3}$ which is not so easy to distinguish from $k^{-5/3}$! Vice versa the spectral slope can change, but the structure remains essentially the same '*yet retaining all the phase information*' (Armi and Flament 1987). Moreover, not only '*the spectral slope alone is inadequate to differentiate between theories*', *alone* it does not correspond to any particular structure(s) in turbulence or it's absence: there is no one-to-one relation between scaling exponents and structure(s) of turbulence. This is true not only of exponents related to Fourier decomposition with its ambiguity (Tennekes 1976), but of many other scaling exponents including those obtained in some wavelet space, SO(3) decomposition and in the physical space—a much overstressed aspect of turbulent flows. Likewise, similar PDFs of *some* quantities can correspond to qualitatively different structure(s) and quantitatively different values of Reynolds number. The emphasis is on *some* quantities like pressure or some other usually (but not necessarily) even order quantities in velocities or their derivatives, since the PDFs of *other* appropriately chosen quantities are sensitive to structure (see below).

As mentioned turbulence possess a number of quasi-Gaussian manifestations. The corresponding statistics belongs to the weakly sensitive to structure(s).

Structure Sensitive Statistics It is noteworthy that—as shown by Hill (1997)—the $-4/5$ Kolmogorov law is more sensitive to the anisotropy, i.e. the third-order statistics (again odd moments), than the second-order statistics. Likewise the structure functions of higher odd orders $S_p^{\|}(r) = \langle(\Delta u_{\|})^p\rangle$ are essentially different from zero, see references in Betchov (1976), Sreenivasan and Antonia (1997), Tsinober (1998b).

Odd Moments and Related PDFs This is an example how structure sensitive statistics can help in looking for the right reasons of measured spectra in the lower mesoscale range (Lindborg 1999). The procedure involves using the third order structure functions which are generally positive in the two dimensional case contrary to the three-dimensional case. Calculations based on wind data from airplane flights, reported in the MOZAIC data set. It is argued that the k^{-3}-range is due to two-dimensional turbulence and can be interpreted as an enstrophy inertial range, while the $k^{-5/3}$-range is probably not due to two-dimensional turbulence and should not be interpreted as a two-dimensional energy inertial range. There is a competing hypothesis that the large scale $-5/3$ range is the spectrum of weakly non-linear internal gravity waves with a forward energy cascade (Van Zandt 1982). A third claim is that the spectral slope in the enstrophy range is more shallow than -3 and is close to $-7/3$ (Tsinober 1995). This range and related anomalous diffusion is explained in terms of the phenomenon of spontaneous breaking of statistical isotropy (rotational and/or reflectional) symmetry—locally and/or globally.

Another example is the demonstration that the small scale structure of a homogeneous turbulent shear flow is essentially *anisotropic* at Reynolds number up to $Re_\lambda \approx 1000$ (Shen and Warhaft 2000); see also Ferchichi and Tavoularis (2000). In order to detect this anisotropy the authors measured the velocity structure functions of third and higher odd orders of both longitudinal and transverse velocity components and corresponding moments of velocity derivatives. In particular, they found a skewness of order 1 of the derivative of the longitudinal velocity in the direction of the mean gradient, which should be very small (or ideally vanish) for a locally isotropic flow. Similar results were obtained in DNS, see references in Tsinober (2009). We should recall that analogous 'misbehavior' of large Reynolds-number turbulence regarding the skewness of temperature fluctuations in the atmospheric boundary layer is known since late sixties (Stewart 1969; Gibson et al. 1970, 1977).

Odd moments such as strain and enstrophy production are obviously of primary importance.

Geometrical Statistics This example shows how conditional sampling based on *geometrical* statistics can help to get insight into the nature of various regions of turbulent flow, e.g. those associated with strong/weak vorticity, strain, various alignments, and other aspects of dynamical importance. The first general aspect is the qualitative difference in the behavior and properties of regions with large enstrophy from strain dominated regions, which is also one of the manifestations of intermittency. Various alignments comprise important simple geometrical characteristics and manifestation of the dynamics and structure of turbulence. For example, there is a distinct qualitative difference between the PDFs of $\cos(\boldsymbol{\omega}, \boldsymbol{\lambda}_i)$ for a real turbulent flow and a random Gaussian velocity field. In the last case, all these PDFs are precisely flat. An example of special dynamical importance is the strict alignment between vorticity, $\boldsymbol{\omega}$, and the vortex stretching vector $W_i \equiv \omega_j s_{ij}$, since the enstrophy production is just their scalar product, $\omega_i \omega_j s_{ij} = \boldsymbol{\omega} \cdot \mathbf{W}$. In real turbulent flows, the PDF of $\cos(\boldsymbol{\omega}, \mathbf{W})$ is strongly asymmetric whereas it is symmetric for a random Gaussian field. It remains essentially positively skewed for any part of the turbulent field, e.g. in the 'weak background' involving whatever definition based on enstrophy, strain, both and/or any other relevant quantity. Thus, contrary to common beliefs, the so called 'background' is not structureless, dynamically not inactive and essentially non-Gaussian, just like the whole flow field or any part of it. The structure of the apparently random 'background' seems to be rather complicated. The previous qualitative observations (mostly from DNS) about the '*little apparent structure in the low intensity component*' or the '*bulk of the volume*' with '*no particular visible structure*' should be interpreted as meaning that no *simple visible* structure has been observed so far in the bulk of the volume in the flow. It is a reflection of our inability to 'see' more intricate aspects of turbulence structure: intricacy and 'randomness' are not synonyms for absence of structure.

Pressure Hessian Some quantities like pressure or other usually (but not necessarily) even order quantities in velocity or their derivatives are less sensitive to structure. The example below present an opposite case

Of special interest is the pressure Hessian $\Pi_{ij} \equiv \frac{\partial^2 p}{\partial x_i \partial x_j}$. Among the general reasons for such an interest is that the pressure Hessian is intimately related to the nonlocality of turbulence in physical space, see references in Tsinober (1998a, 1998b, 2009).

One of the quantities in the present context directly associated with the pressure Hessian is the scalar invariant quantity $\omega_i \omega_j \Pi_{ij}$. It is responsible for the nonlocal effects in the rate of change of enstrophy production $\omega_i \omega_k s_{ik}$. What is special about this quantity, which is of even order in velocity, is that for a Gaussian velocity field $\langle \omega_i \omega_k \Pi_{ij} \rangle_G \equiv 0$, whereas in a real flow it is essentially positive and $\langle \omega_i \omega_k \Pi_{ij} \rangle \sim \frac{1}{3} \langle W^2 \rangle$, where $W_i \equiv \omega_k s_{ik}$ is the vortex stretching vector. Thus interaction between the pressure Hessian and the vorticity is one of the essential features of turbulence structure associated with its nonlocality. It is noteworthy that a similar useful quantity involving strain is non-vanishing for a Gaussian velocity field, $\langle s_{ik} s_{kj} \Pi_{ij} \rangle_G = -\frac{1}{20} \langle \omega^2 \rangle_G^2$.

On Passive Objects and Lagrangian Coherent Structures Above we discussed the dynamical aspects of the problem. The issues of structure(s) in various 'kinematic' issues, like the transport of passive objects (scalars, vectors, etc.), in which Gaussian or other prescribed velocity fields are used rather successfully, can be treated in a similar way as the one described in this section.

As mentioned in Lagrangian setting the dissipative effects are more "influencing" due to strong removal of sweeping effects. Hence stronger deviations from Gaussian statistics in the Lagrangian setting as compared to the Eulerian one just because "inbetween" there is a relation turning even a pure Gaussian velocity field in the Eulerian setting into strongly non-Gaussian one. Hence the so called Lagrangian coherent structures (LCS's) even in pure laminar in the Eulerian setting and pure Gaussian Eulerian velocity field. In this sense the LCS are purely kinematic objects just like the structures in the passive objects evolving in a purely Gaussian velocity field due to the non-linear and non-integrable relation between the Eulerian and Lagrangian fields, for more see Tsinober (2009, p. 300) and Pouransari et al. (2010).

9.2.5 *Structure(s) Versus Scales and Decompositions*

It is natural to ask how meaningful is it to speak about different scales in the context of 'structure(s)' and in what sense, especially when looking at the 'instantaneous' structure(s) of/in turbulence. The known structures indeed possess quite different scales. Vortex filaments/worms—have at least two essentially different scales, their length can be of the order of the integral scale, whereas their cross-section is of the order of Kolmogorov scale. Similarly, the ramp-cliff fronts in the passive scalar fields have a thickness much smaller than the two other scales. This fact is consistent with the observation by Batchelor and Townsend (1949), that *the mean separation of the visible activated regions is comparable with the integral scale of the turbulence, i.e. with the size of the energy-containing eddies.*

It is believed that appropriately chosen decompositions may represent structure(s) of turbulence, e.g. Holmes et al. (1996, 1997). Here again several notes are in order. First, this position depends strongly on what is meant by structure(s). Second, such a possibility is realistic when the flow is dominated by (usually large scale) structures, when many, or practically any reasonable decompositions will do anyhow. And third, structure(s) (and related issues such as geometry) emerging in the 'simplest' case of turbulent flows, in a box with periodic boundary conditions, is(are) are inaccessible via Fourier decomposition, the most natural one in this case.

One of the popular 'decompositions' is into 'coherent structures' and random/dissipative 'background'.[9] This latter is generally considered as structureless and as a kind of passive sink of energy. None is true: the background is not passive at all, it is strongly coupled with the 'coherent structures', and possess lots of it 'own' structure(s).

There is no turbulence without structure(s). Every part (just as the whole) of the turbulent field—including the so-called 'structureless background'—possess structure. Structureless turbulence (or any of its part) contradicts both the experimental evidence and the Navier–Stokes equations. The qualitative observations on the *little apparent structure in the low intensity component* or the *bulk of the volume* with *no particular visible structure should be interpreted as indicating that no simple visible structure* has been observed so far in the bulk of the volume in the flow. It is a reflection of our inability to 'see' more intricate aspects of turbulence structure: intricacy and 'randomness' are not synonyms for absence of structure.

Another kind of decomposition is represented by a latest example attempting to take into account the undeniable structure of the above mentioned "structureless background" by dividing the flow in two characteristic regions: the mentioned above "thin shear layers" occupying a small part of the volume of and the quasi-homogeneous rest (Hunt et al. 2010; Ishihara et al. 2011 and references therein). The assumption of "thin" is necessary in order to employ a kind of RDT approach. Another recent example is the mentioned above by Leung et al. (2012). Apart of

---

[9]An example of a typical statement is represented by the following: *The emergence of collective modes in the form of coherent structures in turbulence amidst the randomness is an intriguing feature, somewhat reminiscent of the mix between the regular "islands" and the "chaotic sea" observed in chaotic, low-dimensional dynamical systems. The coherent structures themselves approximately form a deterministic, low-dimensional dynamical system. However, it seems impossible to eliminate all but finite number degrees of freedom in a turbulent flow—the modes not included form an essential, dissipative background, often referred as an eddy viscosity, that must be included in the description* (Newton and Aref 2003).

problematic nature of such decompositions from the fundamental point, e.g. the claims that the components of some decomposition represent physically meaningful "structures", there are many problems of conceptual and technical nature with what is called 'coherent structures', "thin shear layers", vortex structures, filaments etc., starting from the very beginning of their definition (in fact non-existent or at best vague) and ending with their role in fluid flows both in Eulerian and Lagrangian setting. It is for this reason that *At this stage, this alternative approach (i.e. the 'structural') has not led to a generally applicable quantitative model, neither—for better or worse—has it a major impact on the statistical approaches. Consequently the deterministic viewpoint is neither emphasized nor systematically presented* (Pope 2000). This does not mean that there exists "*generally applicable quantitative model*" based on statistical approaches. It looks that so far Liepmann was correct (but a bit over-optimistic) in his prediction: *Clearly, the exploration of the concept of coherent structure is still on the rise. Turbulence is and will remain the most difficult problem of fluid mechanics, and the past experience suggests that the subsequent fall of interest in the coherent structures is more than likely. The resulting net gain in understanding of turbulence may be less than our expectations of today but will certainly be positive* (Liepmann 1979). Unfortunately, (so far) the *resulting net gain in understanding of turbulence* is far less than was expected in 1979 and on. Nevertheless, though essentially there is no acceptable definition of "coherent structure" the boldest part of the community wonders about "quantification" and even "the dynamical equations for coherent structures to predict their evolution", see e.g. Holmes et al. (1997). Apart of sensible definition of this finite object one needs also a definition of the "incoherent components" not to mention the tools to handle their interaction. On top of this there is a not just technical question on coherent structures of what? It is the right place to remind that the objects termed "coherent structures" as other terms just structures or alike are still elusive, and may appear to be not much different from the Emperors's new Clothes.

The reason for the above statement is as follows.

In dynamical systems, one looks for structure in the *phase space* (Shlesinger 2000; Zaslavsky 1999), since it is relatively 'easy' due to low dimensional nature of the problems involved. In turbulence nothing is known about its properties in the corresponding very high dimensional phase space.[10] Therefore, it is common to look for structure in the *physical space* with the hope that the structure(s) of turbulence—as we observe it in *physical space*—is (are) the manifestation of the generic structural properties of mathematical objects *in phase space*, which are called attractors and which are invariant in some sense. In other words, the structure(s) is (are) assumed to be 'built in' in the turbulence independently of its (their) origin. The problem is that due to very high dimension and complex behavior of turbulent flows and structure of the underlying attractors *one may never be able to realistically determine the fine-scale structure and dynamic details of attractors of even moderate dimension... The theoretical tools that characterize attractors of moderate or large*

[10]Hopf (1948) conjectured that the underlying attractor is finite dimensional due to presence of viscosity.

dimensions in terms of the modest amounts of information gleaned from trajectories [i.e., particular solutions] *... do not exist... they are more likely to be* ***probabilistic than geometric in nature*** (Guckenheimer 1986). Therefore, it is indeed unlikely that one can succeed in hunting individual structures of finite dimensions using low-dimensional tools not to mention isosurfacing based on one parameter since evolution of finite objects is not low dimensional. The remaining question is what does one see in reality named as "structure", "coherent structure" and so on. This question deserves far more serious attention beyond weakly founded speculations.

At present the dynamical systems community advocates an alternative approach to turbulence, based on recently found simple invariant solutions and connecting orbits in Navier–Stokes flows (Cvitanović and Gibson 2010; Kawahara et al. 2012). This requires handling of ODEs systems with a large number degrees of freedom for pretty moderate Reynolds numbers, typically larger 10^5 for $Re \sim 10^2$, which hardly can be qualified as low-dimensional. This means that even here one cannot avoid statistics. However, though the system of these equations seems to be not far easier to solve than the full time-dependent three-dimensional Navier–Stokes equations there is an important hope and even promise to "get into" what von Neumann wrote about in 1949: *nothing less than a thorough understanding of the* [global behavior of the] *system of all their solutions would seem to be adequate to elucidate the phenomenon of turbulence... There is probably no such thing as a most favored or most relevant, turbulent solution. Instead, the turbulent solutions represent an ensemble of statistical properties, which they share, and which alone constitute the essential and physically reproducible traits of turbulence.* The results obtained so far are significant for a theoretical description of transition to turbulence. However, the claim that the same is true of "also fully turbulent flow" seems to be a bit premature even for low Reynolds numbers $\leq 10^3$ as long as one talks about things like "resembling the spatially coherent objects found in the near-wall region of true turbulent flows", having "the potential to represent coherent structures", "simple invariant solutions could represent turbulence dynamics, whereas the simple solutions themselves would represent coherent structures embedded in a turbulent state", reminiscence of things observed in DNS as "regeneration cycle in the buffer layer", etc. All this for low Reynolds numbers with considerably lower capability than DNS of NSE. The bottom line is that one is tempted to ask the question by Cvitanović and Gibson (2010) *Should this be called 'turbulence?'*. Resemblance is far less than necessary and definitely not sufficient.

Part IV
Epilogue

As mentioned at the very beginning there is far more to say about the difficulties rather than achievements when it comes to the basic aspects of the problem. we reiterate some main points with a bit different accents which are followed by some remarks on what next.

Chapter 10
On the Status

Abstract As mentioned at the very beginning there is far more to say about the difficulties rather than achievements when it comes to the basic aspects of the problem. All ideas/ theories, etc. proposed so far did fail falling into category of misconceptions and/or ill defined concepts, see Chap. 9 in Tsinober (An informal conceptual introduction to turbulence, 2009). We reiterate some main points with different more general accents. These are followed by some remarks on what next.

- Cascade (versus decompositions), inertial range and its intermittency and 'anomalous scaling', 'hierarchical structure' of turbulence and their numerous representations, rather than explanation, by things like breakdown coefficients/multipliers (Novikov 1974, 1990a) or equivalently multi-fractals (Frisch 1995) and the so-called 'hierarchical symmetry' (She and Zhang 2009). The main problems are due to misinterpretations of observations such as in the case of the famous by Richardson poem 1922 among many subsequent and more general problems in using experimental data; almost infinite belief in some hypotheses as in case of existence of inertial range and, more generally, excessive belief in locality properties of turbulence; the view that turbulent flows are hierarchical, which underlies the concept of the cascade, though convenient, is more a reflection of the unavoidable (due to the nonlinear nature of the problem) hierarchical structure of models of turbulence and/or decompositions rather than reality; similarly view that turbulence structure(s) is(are) simple in some sense and that turbulence can be represented as a collection of simple objects only seems to be a nice illusion which, unfortunately, has little to do with reality. It seems somewhat wishfully naive to expect that such a complicated phenomenon like turbulence can merely be described in terms of collections of only such 'simple' and weakly interacting objects. We remind again that structure—whatever this means—can be very simple and complex in the same flow depending on the field of interest, see Fig. 1.3. In particular, flow visualizations used for studying the structure of dynamical fields (velocity, vorticity, etc.) of turbulent flows may be quite misleading, making the question "what do we see?" extremely nontrivial. The commonly used isosurfacing/thresholding cannot be expected to correspond to generic aspects of turbulence structure(s) and objects in question many of which still await to be found and properly defined. There is no turbulence without structure(s). Every

A. Tsinober, *The Essence of Turbulence as a Physical Phenomenon*,
DOI 10.1007/978-94-007-7180-2_10,

part (just as the whole) of the turbulent field—including the so-called 'structureless background'—possess structure. Structureless turbulence (or any of its part) contradicts both the experimental evidence and the Navier–Stokes equations. The qualitative observations on the *little apparent structure in the low intensity component* or the *bulk of the volume* with *no particular visible structure* should be interpreted as indicating that no *simple visible* structure has been observed so far in the bulk of the volume in the flow. It is a reflection of our inability to 'see' more intricate aspects of turbulence structure: intricacy and 'randomness' are not synonyms for absence of structure.

- Origins of vorticity amplification, prevalence of vortex stretching and high rate of dissipation in turbulent motion (Taylor 1938a, 1938b). The main problems are due a misconception that vorticity amplification, which is a dynamical process, is due to the same mechanism as stretching of material lines, which is a kinematical process as with other passive objects, see Chap. 9 in Tsinober (2009); misinterpretation of the prevalence of vortex stretching—the enhanced dissipation is due to strain production and prevalence of compression in this process. It is the strain production which plays the role of an engine producing the whole field of velocity derivatives, both itself and the vorticity. It is of special importance on paradigmatic level that it is the strain production which is responsible for the finite overall dissipation at (presumably) any however large Reynolds numbers. An important aspect is that the field of strain is efficient in the above two missions only with the aid of vorticity, i.e. only if the flow is rotational. There is a conceptual and qualitative difference between the nonlinear interaction between vorticity and strain, e.g. $\omega_i \omega_j s_{ij}$ and the self-amplification of the field of strain i.e., $s_{ij} s_{jk} s_{ki}$, which is a specific feature of the dynamics of turbulence having no counterpart (more precisely analogous—not more) in the behavior of passive and also active objects. This process, i.e., $s_{ij} s_{jk} s_{ki}$ is local in contrast to $\omega_i \omega_j s_{ij}$, as the field of vorticity and strain are related nonlocality.
- The problem of closure, eddy viscosity, low dimensional description, LES and similar and even "the dynamical equations of coherent structures". This is set of problems comprising an everlasting dream to be able to, e.g. obtain the mean properties and large scale properties of turbulence without solving directly the NSE. So no wonder *that most of the theoretical work on the dynamics of turbulence has been devoted (and still is devoted) to ways of overcoming the difficulties associated with the closure problem* (Monin and Yaglom 1971, p. 9). These difficulties have not been overcome and it does not seem that this will happen in the near future if at all. Moreover, not everybody in the community will agree to qualify this work as really theoretical. Nevertheless the common and even massive practice is to employ some "parameterization", e.g. representing *the unresolved scales in climate models by imagining an ensemble of sub-grid processes in approximate secular equilibrium with the resolved flow* (Palmer 2005), see also Palmer and Williams (2008). The use of such approaches—all being a kind of low dimensional description—is successful and even adequate in the narrow sense as a (semi) empirical tool. However, from the basic point of view the dream about the low-D description seems to be ephemeral—there cannot be adequate

low-D description of turbulence—this would be a major oversimplification of the complex interaction of large/resolved and small/underresolved scales including the direct and bi-directional, i.e. non-locality. This concerns also a set of issues on the adequacy of quasi-two-dimensional approximations, e.g. models of large scale atmospheric motions with poor (if any) understanding of their interaction (direct and bidirectional) with the smaller scale three-dimensional turbulent flow.

Indeed, looking at the equations for the small/unresolved scales it is straightforward to realize that the small/unresolved scales depend on the large/resolved scales via nonlinear space and history-dependent functionals, i.e. essentially non-local both spatially and temporally. So it is unlikely—and there is accumulating evidence for this—that relations between them (such as "energy flux", but not only) would be approximately local in contradiction to K41a hypotheses and surprisingly numerous attempts to support their validity.

Among the strong arguments for this is that the low-D description is killing an important part of turbulence physics residing in the small scales and in particular those associated with the dissipative and rotational properties of turbulence—the two properties of distinct paradigmatic nature and importance. Looking for the right reasons the unresolved scales cannot be represented by *simple formulae relating them to rest of the flow* due to non local properties of turbulence among other reasons. In other words, in the low-D approaches the rotational and dissipative properties do not seem to be considered as belonging to the essential dynamics due to the very nature of low-D approaches: *The basic question* (which usually is not asked) *concerning statistical description is whether such complex behavior permits a closed representation that is simple enough to be tractable and insightful but powerful enough to be faithful to the essential dynamics* (Kraichnan and Chen 1989). Thus from fundamental point of view a fluid flow which is adequately represented by a low-dimensional system is not turbulent—a kind of definition of 'non-turbulence'.

Two additional remarks.

The first concerns use of statistical methods and possible alternative. There is an essential difference between the enforced necessity to employ statistical methods in view absence of other methods and tools so far and the impossibility in principle (Monin and Yaglom 1971), to study turbulence via other approaches. This is especially discouraging all attempts to get into more than just "en masse". Also such a standpoint means that there is not much to be expected as concerns the essence of turbulence using statistical methods only. In reality other approaches are being used in some special cases as described in the text. At present the dynamical systems community advocates an alternative approach to turbulence. There is some hope and possibly promise to "get into" what von Neumann wrote about in 1949 at least at low Reynolds numbers.

A final remark here is about the ambiguity of turbulence language with imperfect/inadequate terminology which is characteristic of an underdeveloped branch of science. This is related to ill defined concepts either, etc. The terminological problem is divided roughly in two. One is about more or less specific like terms as "scales" and "structures" and also some buzzwords, the other concerns more general terms as "relevant" and "significant".

10.1 What Next

As concerns genuine theory it is hard to imagine clear optimistic expectations. Indeed, along with absence of genuine theory there is no consensus on what is (are) the problem(s) of turbulence and what would constitute its (their) solution; there is no consensus on what are the main difficulties and why turbulence is so impossibly difficult: almost every aspect of turbulence is controversial, which by itself is one of the greatest difficulties. Neither is there agreement on what constitutes understanding which can be brought only by a genuine theory: just like no sophisticated experiment (laboratory or DNS[1]) by itself does not bring understanding, neither does modeling of whatever sophistication, see references and quotations in the Appendix. Judging by previous experience there is no basis for optimism as concerns all known theoretical approaches definitely including statistical methods due to their inherent limitations.

We remind that this is not to claim absence of theory(ies). On the contrary, there are plenty, many with qualitatively different and even contradictory premises but all agreeing well with some experimental data and even claiming rigor—people attempting to find it are in danger of a non-trivial waste of time—but not necessarily for the right reasons and not based on first principles with few exceptions having no direct bearing on the Navier–Stokes equations.

There is a need for a genuine physical theory with some non-trivial understanding of the physics taking into account the nonlocal properties of turbulence—no theory based on some "locality", "small parameters", etc., has little chance to succeed if any. There seem to exist no small parameters, except of the genuinely small parameter introduced by Saffman (1978), which he called information density, ϵ_I, and defined as the ratio, S/N, in the literature, with $S =$ signal (*understanding*), and $N =$ noise (*mountains of publications*).

In view of the general theoretical failure, i.e. absence of theory based on first principles a number of issues of special concern are about the relation(s) of what is called theory today in turbulence research and observations/experiments. These include the use of the factual information as concerns fundamental aspects as it stands now the experimental 'confirmation' of a 'theory' became meaningless. The above concern includes also the issue of non-trivial resources (not just money) needed for serious progress. Special care is needed as there is no theory to guide the observations.

---

[1] *Progress in numerical calculation brings not only great good but also awkward questions about the role of the human mind... The problem of formulating rules and extracting ideas from vast masses of computational or experimental results remains a matter for our brains, our minds* (Zeldovich 1979).

There are essential differences between physical and numerical experiments. If underresolved the former still provide correct information which is a problem with the latter especially as concerns numerical errors sometimes interpreted as genuine chaos.

There exist a great variety of observations in a great variety of conditions. The problem is that the theoretical community did not express much interest in most because almost if not all these observations did not match the requirements of a variety of idealized concepts and "simplifications" such as homogeneity and anisotropy both local and global, scaling and other symmetries, universality, periodic BC's, very large Reynolds numbers, etc.

It is becoming clear that such a matching is impossible as the idealizations are far more different than conditions corresponding to all observations. Thus the turbulence community is in danger to stay divided forever between experimentalists who observe what cannot be explained and theoreticians who try to explain what can not be observed, and there will be more attempts to replace explanation with mere description. The existing theoretical "frame" of idealizations is too narrow mainly due to problems associated with nonlocal effects. Insistence on such a "frame" results in misinterpretation of experimental data and irrelevant models such as in the case of "anomalous scaling" and intermittency in the inertial range—a nonexistent object. In other words, due to nonlocal effects—quite a bit of manifestations—one cannot generally assume that approximately homogeneous (isotropic) regions in a bounded region in flows, which are otherwise non-homogeneous, have the same properties as those globally homogeneous, though it is done this pretty frequently. This is also true of local homogeneity and isotropy.

The issue of interpretation and validation refers not just to "theories", which anyway are not in existence, but in the first place to the factual information. The main point here is that the right results should be interpreted and related to the right reasons at least as concerns fundamental studies. The correspondence with experimental results may and mostly does occur for the wrong reasons, i.e. this correspondence is at best a necessary condition. One of the problems is misinterpretation, which is "aided" by the extremely complex nature of the problem. Another acute problem is that most of the theories claim explanations of some specific aspects of observations being in fact mere descriptions of these same particular observations.

An example of such a position is found in Monin and Yaglom (1971, p. 21): *...we have avoided introducing experimental results which have no theoretical explanation and which do not serve as a basis for some definite theoretical deductions, even if these data are in themselves very interesting or practically important.* This is because *one of the principal incentives for writing this book was a desire to summarize the development of the idea of a universal local structure in any turbulent flow for sufficiently large Reynolds number.* In other words, they looked for confirmation of something very important as everybody does. Unfortunately, just like stated by Liepmann (1979), digging a bit deeper into the observations evidence proved that what Monin and Yaglom were seeking appeared as problematic as all previous and subsequent ideas though far more popular. However, quoting Saffman (1978), *It must of course be kept in mind that the achievements or intrinsic value of a theory is not decided by democratic means. The majority view in science is not necessarily the right one.* This stands is drastic contradiction with the modern 'culture' of evaluation such as 'ratings' and similar bureaucratic 'inventions', which contributed significantly to the alarming overabundance and continuing major over-production of publications. Another aspect is about the attitude of modelers: *Whenever they fail in their predictions, scientists tend to blame the poor accuracy of the observations, the lack of computer power and the inadequate parametrization in their numerical models, rather than their own lack of skill in computing the accuracy that can be obtained with present resources* (Tennekes 1993).

The bottom line is that a narrow mathematical, theoretical, observational, etc. standpoint is outstandingly and especially harmful for basic research in turbulence. The exception seems to be a physical standpoint understood in as broad sense as one can imagine.

10.2 What to Do

The answer so far is to move along the von Neumann's suggestion with the hope to break the deadlock, but not only by extensive well planned computational efforts. It became obvious that DNS is not as powerful as claimed for quite a while, von Neumann was too optimistic especially as concerns large Reynolds numbers. Physical experiments are badly needed especially at large Reynolds numbers as advised by Kolmogorov. Among the observations among the most important are those associated with the rotational and dissipative nature of turbulence and its nonlocal properties. Though there is no theory to guide the observations we are lucky to be in possession of the NSE. It is true that there is little substantial theoretical use of NSE in turbulence, since there is almost no way to use them explicitly in theoretical approaches, e.g. by solving them by 'hand'. However, there are several ways to do this implicitly, i.e. by indirect use of NSE and their consequences. For example, looking at the NSE and their consequences such as those for vorticity strain, helicity and some other themselves enables us to recognize the dynamically important quantities and physical processes involved. In other words, NSE and their consequences tell us at least in part what quantities and relations should be studied, see e.g. pp. 366–367 in Tsinober (2009) for the equations of evolution of enstrophy and strain production revealing the invariant quantities $\omega_i \omega_j \Pi_{ij}$ and $s_{ik} s_{kj} \Pi_{ij}$. Another example is the equation for the "energy" of strain s^2 discussed in Chap. 7 above which by itself allowed to arrive to the unequivocal conclusion about the key role of the strain production as concerns the issues of self-amplification of velocity derivatives rather that the enstrophy production.

One of the problems is that any serious planning of experiments (numerical, laboratory, field) cannot be undertaken without reasonable a priory basic understanding. The lack of such comprises a real deadlock in turbulence research. So the success of basic research depends strongly on scientific curiosity and those who are not afraid of real adventure. In spite and because of the state of matters in the field it is very much rewarding to dwell into this mysterious world and the hope is on intelligent, clever and courageous young people.

An important question is what not to do. As the reader already realized from the text the list is not short. In any case one has to be definitely extremely careful in "validation" of theories, as the latter appear pretty fast out of existence, unless the "new idea" looks enough trustworthy and attractive. Too little is still known about the properties of real turbulence. So as concerns basic issues it seems not justified to put too much pretty futile efforts in it's modeling which mostly is mimicking it without much understanding. Hence the necessity to follow the advice of those cited at the beginning of Part I. The observational aspect is not that trivial in such a highly-dimensional system as turbulence if one does not stay with a hopeless choice of one velocity component.

The future of basic research depends strongly on questions of paradigmatic nature independently and as contrasted to pragmatic needs in applications. It is far more difficult close to impossible to ask the right questions to promote the basic research. However, this contrast is only short-term and as shows history of natural sciences disappears in the long run. The contrast may extend for a long time in case when "science" is pursued solely/mainly for the purposes of making better weapons and neater gadgets and, of course, money. But in such a case it is destined to degeneration as science: ***There are no such things as applied sciences, only applications of science*** (Louis Paster 1872, Address 11 Sept 1872, Comptes Rendus des Travaux du Congress Viticole et Sericole de Lyon, 9–14 Septembre, 1882, p. 49).

Chapter 11
Appendix. Essential Quotations

Abstract Along with a relatively limited number of intext citations we bring a small collection of citations in this Appendix, more is found in Tsinober (An informal conceptual introduction to turbulence, 2009). In this reference one of the aims was an extensive treatment of the dialog in the turbulence community with an emphasis on problems of a conceptual nature.

Here we put more emphasis on the pretty early recognition in the community of the fundamental difficulties as concerns the theoretical aspects of the problem and as a consequence the absence of theory based on first principles and inadequate tools to handle both the problem and the phenomenon of turbulence. This rather peculiar state of matters exists along with a set of deterministic differential equations, the Navier–Stokes equations probably containing all of turbulence, so that most of our knowledge about turbulence comes from observations and experiments, laboratory, field and later numerical, which is unfortunate as theory is supposed to guide and gives meaning to observation.

Along with absence of genuine theory there is no consensus on what is (are) the problem(s) of turbulence and what would constitute its (their) solution. Neither is there agreement on what constitutes understanding which can be brought only by a genuine theory.

This state of matters was understood long ago by many outstanding people, Kolmogorov, von Neumann, Wiener and many other as quoted below. Related issues concern the multitude of "approaches" and the continuing diversity of opinions on what is important, what are the main questions and similar.

11.1 To Preface

11.1.1 On Absence of Genuine Theory

• *My overall impression of the Symposium is that no really new and important ideas have been presented... I think we must admit that little new theory has been put before us.* Batchelor 1959 Some reflections on the theoretical problems raised at the Symposium, Proceedings of a Symposium on Atmospheric diffusion and air pollution, editors F.N. Frenkiel and P.A. Sheppard, Advances in Geophysics, **6**, 449–452.

A. Tsinober, *The Essence of Turbulence as a Physical Phenomenon*,
DOI 10.1007/978-94-007-7180-2_11,

• *Formal mathematical investigations have produced remarkably little value... A number of general procedures for calculation of various dynamical aspects of homogeneous turbulence have been devised, but none of them impresses me as being likely either to advance our understanding of turbulence or to achieve results on which we can place reliance... The universal similarity theory of the small-scale components of the motion stands out in this rather grey picture as a valuable contribution.* Batchelor G.K. 1962 The dynamics of homogeneous turbulence: introductory remarks, In A. Favre, editor, Mécanique de la Turbulence, Colloques Interntionaux du CNRS, No. 108, Marseille, 28 août–2 septembre 1961, p. 96.

• *Turbulence was probably invented by the Devil on the seventh day of Creation when the Good Lord wasn't looking.* P. Bradshaw 1994 Experiments in Fluids, **16**, 203.

• *Turbulence is the last great unsolved problem of classical physics. Remarks of this sort have been variously attributed to Sommerfeld, Einstein, and* ***Feynman****, although no one seems to know precise references, and searches of some likely sources have been unproductive. Of course, the allegation is a matter of fact, not much in need of support by a quotation from a distinguished author. However, it would be interesting to know when the matter was first recognized.* P.J. Holmes, G. Berkooz and J.L. Lumley 1996 Turbulence, coherent structures, dynamical systems and symmetry, Cambridge University Press.

• *As a doctorate I proposed to Heisenberg no theme from Spectroscopy but the difficult problem of Turbulence, in the hope, that WENN IRGENDEINER (if anybody), would solve this problem. However, the problem is until now not solved.* A. Sommerfeld 1942, Scientia, Nov./Dez. 1942.

• *I soon understood that there was little hope of developing a pure, closed theory, and because of absence of such a theory the investigation must be based on hypotheses obtained on processing experimental data,* Kolmogorov A.N. 1985, in notes preceding the papers on turbulence in the first volume of his selected papers; English translation, Selected works of A.N. Kolmogorov, **I**, ed. Tikhomirov, p. 487, Kluwer.

• *...the essential mathematical complications of the subject were only disclosed by actual experience with the physical counterparts of these equations... The entire experience with the subject indicates that the purely analytical approach is beset with difficulties, which at this moment are still prohibitive. The reason for this is probably as was indicated above: That our intuitive relationship to the subject is still to loose—not having succeeded at anything like deep mathematical penetration in any part of the subject, we are still quite disoriented as to the relevant factors, and as to the proper analytical machinery to be used. Under these conditions there might be some hope to "break the deadlock" by extensive, but well planned computational efforts.* J. von Neumann 1949 Recent theories of turbulence—A report to Office of Naval Research. Collected works, **6** (1963), pp. 468–469, ed. Taub., A.H., Pergamon. Note the 'bias' of a mathematician/theoretician: equations first, but cf. the statement by Kolmogorov 1985.

• *It remains to call attention to the chief outstanding difficulty (i.e. turbulence) of our subject.* H. Lamb 1927 Hydrodynamics, p. 651.

• *I am an old man now, and when I die and go to Heaven there are two matters on which I hope for enlightenment. One is quantum electrodynamics and the other is the turbulent motion of fluids. And about the former I am rather optimistic.* Sir Horace Lamb as quoted by S. Goldstein 1969Goldstein (1969), ARFM, **1**, 23.

• *Turbulence is the graveyard of theories,* H.W. Liepmann, 1997 A brief history of boundary layer structure research, in Self-sustaining mechanisms of wall turbulence, editor R.L. Panton, p. 4, Comp. Mech. Publ.

• *It is much easier to present nice rational linear analysis than it is to wade into the morass that is our understanding of turbulence dynamics. With the analysis, professor and students feel more comfortable; even the reputation of turbulence may be improved, since the students will find it not as bad as they had expected. A discussion of turbulence dynamics would create only anxiety and a perception that the field is put together out of folklore and arm waving* Lumley, J.L. 1987 Review of 'Turbulence and random processes in fluid mechanics' by M.T. Landahl and E. Mollo-Christensen, J. Fluid Mech., **183**, 566–567.

Even after 100 years turbulence studies are still in their infancy. We do have a crude practical working understanding of many turbulence phenomena but certainly nothing approaching comprehensive theory and nothing that will provide predictions of an accuracy demanded by designers. Lumley, J.L. and Yaglom, A.M. 2001 A century of turbulence, *Flow, Turbulence and Combustion*, **66**, 241–286.

• *It is at this point that the study of turbulence does prove to be an exception: the applied physics involvement is almost completely absent. In view of the extraordinary practical importance of turbulence…, this is quite astonishing. Yet the reason for such apparent neglect is easily found. Quite simply the fundamental problems of turbulence are still unresolved.* D. McComb 1990 The physics of turbulence, Oxford Univ. Press, p. vii.

• *While the experimental techniques that have been invaluable in understanding phase transitions promise to be very useful in the study of hydrodynamic phenomena, I suspect that the recent addition to our theoretical arsenal may be less effective than many had hoped* Martin, P.C. 1976 The onset of turbulence: A review of recent developments in theory and experiment, in: L.Pál and P. Szépfalusy, editors, Statistical physics: Proceedings of the International Conference, Budapest, 5–29 August, 1975, pp. 69–96. Amsterdam: North-Holland.

• *…a fundamental theoretical understanding is still lacking.* M. Nelkin, 1994 Universality and scaling in fully developed turbulence, Adv. Phys., **43**, 143.

• *The existence of an asymptotic statistical state is strongly suggested experimentally, in the sense that reproducible statistical results are obtained. However, physical plausibility aside, it is embarrassing that such an important feature of turbulence as its statistical stability should remain mathematically unresolved, but such is the nature of the subject.* Orszag, S.A. 1977, Lectures on the statistical theory of turbulence, in: R. Balian and J.-L. Peube, editors, *Fluid Dynamics*, pp. 235–374, Gordon and Breach.

• *we should not altogether neglect the possibility that there is no such thing as 'turbulence'. That is to say, it is not meaningful to talk of the properties of a turbulent flow independently of the physical situation in which it arises. In searching for a theory of turbulence, perhaps we are looking for a chimera,* P.G. Saffman 1978 Problems and progress in the theory of turbulence, Lect Notes in Phys., **76** (II), 276.

• *One of the things that I always found troubling in the study of the problem of turbulence is that I am not quite sure what the theoretical turbulence problem actually is... One reason I think we have so much difficulty in solving it, is that we are not really sure what it is... I just cannot think of anything where a genuine prediction for the dynamics of turbulent flow has been confirmed by an experiment. So we have a big vast empty field.* P.G. Saffman 1991 in The Global Geometry of Turbulence, NATO ASI Ser. B 268, ed. J. Jimenez, pp. 348, 349, Plenum.

• *In contrast to this experimental cornucopia, theory can offer only a few crumbs.* Siggia, E. 1994 High Rayleigh number convection, Ann. Rev. Fluid Mech., **26**, 137–168.

• *In spite of a huge number of papers and a large amount of research on turbulence, it remains an unsolved problem left for future generations.* Sinai, Ya.G. 1999 Mathematical problems of turbulence, Physica, **A263**, 565–566.

• *...the absence of a sound theory is one of the most disturbing aspects of the turbulence syndrome.* R.W. Stewart, 1969, Turbulence Nat. Committee for Fluid Motion Films, dist. Encyclopedia Britannica Educational Corp.

• *Sometimes experiments provide us with so beautiful and clear results that it is a shame on theorists that they cannot interpret them,* Yudovich, 2003.

• *It has been realized since the beginning that the problem of turbulence is a statistical problem; that is a problem in which we study instead of the motion of a given system, the distribution of motions in a family of systems... It has not, however, been adequately realized just what has to be assumed in a statistical theory of turbulence,* Wiener, N. 1938 Homogeneous chaos, *American Journal of Mathematics*, **60**, 897–936.

11.2 To Chap. 1

11.2.1 On Multitude of "Approaches"; for More See Also Chaps. 3 and 9 in Tsinober (2009)

• *Kolmogorov's ideas on the experimentalist's difficulties in distinguishing between quasi-periodic systems with many basic frequencies and genuinely chaotic systems have not yet been formalized.* Arnold 1991.

• *From the point of view of theoretical physics, turbulence is a classical field theory, out of equilibrium and in a strong coupling regime*, Announcement by Chinese Academy of Sciences of a programme New Directions in Turbulence, held at Bejing, Mar 12 to Apr 20, 2012.

• An example from the mathematical community by Foias et al. 2001: *The word turbulence has different meanings to different people,which indicates that turbulence is a complex and multifaceted phenomenon. For mathematicians, outstanding problems revolve around the Navier–Stokes equations (such as wellposedness and low-viscosity behavior,especially in the presence of walls or singular vortices). For physicists, major questions include ergodicity and statistical behavior as related to statistical mechanics of turbulence. Engineers would like responses to questions simple to articulate but amazingly difficult to answer: What are the heat transfer properties of a turbulent flow? What are the forces applied by a fluid to its boundary (be it a pipe or an airfoil)? To others pursuing the dynamical system approach, of interest is the large time behavior of the flow. Another ambitious question for engineers is the control of turbulence (to either reduce or enhance it), which is already within reach. Finally, a major goal in turbulence research—of interest to all and toward which progress is constantly made—is trustworthy and reliable computation of turbulent flows.*

• *In the present-day statistical fluid mechanics, it is always implied that the fluid mechanical fields of a turbulent flow are random fields in the sense used in probability theory.* Monin and Yaglom, 1971, pp. 3–4, 7.

• *It is natural to assume that in a turbulent flow... the... fluid dynamic variables, will be random fields* (Monin and Yaglom 1971, p. 214).

• *From the very beginning it was clear that the theory of random functions of many variables (random fields), whose development only started at that time, must be the underlying mathematical technique.* A.N. Kolmogorov, 1985 in notes preceding the papers on turbulence in the first volume of his selected papers, English translation, Tikhomirov, 1991, p. 487.

• A list of interest is from the engineering community by Lumley (1990): *Turbulence is rent by factionalism. Traditional approaches in the field are under attack, and one hears intemperate statements against long time averaging, Reynolds decomposition and so forth. Some of these are reminiscent of the Einstein–Heisenberg controversy over quantum mechanics, and smack of a mistrust of any statistical approach. Coherent structure people sound like The Emperors's new Clothes when they say that all turbulent flows consist primarily of coherent structures, in the face of visual evidence to the contrary. Dynamical systems theory people are sure that turbulence is chaos. Simulators have convinced many that we will be able to compute anything within a decade... The card-carrying physicists dismiss everything that has been done on turbulence from Osborne Reynolds until the last decade. Cellular Automata were hailed on their appearance as the answer to a maidens prayer, so far as turbulence was concerned.* Lumley 1990, as quoted by Cantwell, B.J. 1990 Future directions in turbulence research and the role of organized motion, in J.L. Lumley (ed)., *Whither turbulence?,* pp. 97–131, Springer.

• *I think that the k-space decomposition does actually obscure the physics.* Moffatt, H.K. 1990 in: J.L. Lumley, editor, Whither turbulence? Turbulence at the crossroads, Springer, Berlin, p. 296.

• *... one may never be able to realistically determine the fine-scale structure and dynamical details of attractors of even moderate dimension. The theoretical tools that characterize attractors of moderate or large dimensions in terms of the modest amounts of information gleaned from trajectories* [i.e. particular solutions]*... do not exist... they are more likely to be probabilistic than geometric in nature.* Guckenheimer, J. 1986 Strange attractors in fluids: another view, Ann. Rev. Fluid. Mech., **18,** 15–31.

11.3 To Epilogue

11.3.1 On the Continuing Diversity of Opinions on What Is Important, What Are the Main Questions and Related

• *...it is amazing how many different, nearly orthogonal, points of view there are about a phenomenon which is governed by Newton's innocent-looking, linear second law of motion, with a little help or hindrance from viscosity.* Bradshaw 2003 Review of "Turbulent Flow: Analysis, Measurement and Prediction" by P.S. Bernard & J.M. Wallace. Wiley, 2002, J. Fluid Mech., **478**, 344–345.

• *A principal objective of any theory of fluid motion is the prediction of the spread of matter or "tracer" within the fluid*, Bennet, A. 2006 Lagrangian fluid dynamics, CUP.

• *The main ambition of the modern theory of turbulence is to explain the physical mechanisms of intermittency and anomalous scaling in different physical systems, and to understand what is really universal in the inertial-interval statistics.* Falkovich, G., Gawedzki, K., and Vegassola, M. 2001 Particles and fields in fluid turbulence, Rev. Mod. Phys., **73**, 913–975.

• *If we assume as a basic starting point in every theory of turbulence its representation in terms of spectral coefficients, statistical or physical averages, or more generally simple objects conditionally extracted by a weak background, turbulence modeling could be defined reductively, as the art of writing the equations that produce directly such quantities.* Germano, M. (1999) Basic issues of turbulence modeling, in: A. Gyr, W. Kinzelbach and A. Tsinober, editors, Fundamental Problematic Issues in Turbulence, 213–219, Birkhäuser.

• *The separation of scales is a central problem in turbulence.* Germano 2012 The simplest decomposition of a turbulent field, Physica **D 241**, 284–287.

• *A property of turbulent motion is that the boundary conditions do not suffice to determine the detailed flow field but only average or mean properties. For example, pipe flow or the flow behind a grid in a wind tunnel at large Reynolds number is such that it is impossible to determine from the equations of motion the detailed flow at any instant. The true aim of turbulence theory is to predict the mean properties and their dependence on the boundary conditions.* Saffman P.G., 1968 Lectures on homogeneous turbulence, in: N.J. Zabusky, editor, Topics in nonlinear physics, pp. 485–614, Springer.

• *An early goal of the statistical theory of turbulence was to obtain a finite, closed set of equations for average quantities, including the mean velocity arid the energy spectrum. That goal is now viewed to be unrealistic. The goal nowadays is to reduce to a manageable number the many degrees of freedom necessary to describe the flow, to determine the equations governing the dynamics of the reduced degrees of freedom, and to solve those equations analytically or numerically to calculate fundamental quantities that characterize the flow. Thus a theory may treat all or only some of the degrees of freedom statistically.* Frisch, U. and Orszag, S.A. 1990 Turbulence: challenges for theory and experiment, Phys. Today, **43**, 24–32.

• *Herein lies the central obstacle of the entire theory of turbulence known as the closure problem,* Jovanović, J. 2004 The statistical dynamics of turbulence, Springer.

When coupled to the basic conservation laws of fluid flow, such treatment, however, leads to an unclosed system of equations: a consequence reasoned, in the scientific community, the closure problem. This is the central and still unresolved issue of turbulence which emphasizes its chief peculiarity: our inability to do reliable predictions even on the global flow behavior. The equations of motion for this covariance contain third order moments of the velocity field, the equations of motion for the third-order moments contain fourth-order moments, and so forth, ad infinitum. A central goal of turbulence theory is the closing of this infinite chain of coupled equations into a determinate set containing only moments below some finite order. Kraichnan 1957 The structure of isotropic turbulence at very high Reynolds numbers, J. Fluid Mech., **5**, 497–543.

• *The central problem of turbulence is to find the analog of the Gibbs distribution for the energy cascade. The mathematical formulation of this problem is amazingly simple. In the inertial range we can neglect the viscosity and forcing and study the Eulerian dynamics of an ideal incompressible fluid.* Migdal, A.A. 1995 Turbulence as Statistics of Vortex Cells, in Mineev, V.P., editor, The first Landau Institute Summer School, 1993, Gordon and Breach, pp. 178–204.

• Following is a set of questions distributed by John Lumley at the meeting Turbulence at the Crossroads Whither Turbulence? Held at Conell University, Ithaca, NY, March 22–24, 1989.

1. *What are the advantages and disadvantages of long-time averaging (Reynolds averaging)? Is this a technique that is intellectually bankrupt, that systematically excludes certain information, that has mislead us over the years, or is it still viable if used judiciously?*

2. *Where do we stand on statistical approaches generally? Is it a problem that mean quantities seldom occur, or that the things being described statistically have considerable spatial structure? How do we feel about statistical approaches close to critical values of parameters, when the flow is relatively ordered?*

3. *What is our position on coherent structures? Do most turbulent flows consist of more- or -less organized structures, or is there a large range of organization/disorganization in flows with different etiology? How can we use this information?*

4. *What role do we feel Dynamical Systems Theory can play in turbulence in open flow systems? What is the evidence for strange attractors in fluid turbulence? Are there Reynolds/Rayleigh/Richardson number limitations on the utility of these approaches?*

5. *Where do we stand on large eddy simulation and exact simulation vis a vis computer development? Is it only a matter of a couple of decades until we will be able to calculate everything, and hence we can soon abandon all modeling and experiment, or are some things still going to be out of reach? Which things?*

6. *How do we feel about modeling? Granted that .there has been a lot of bad modeling done, there are more fundamental issues: Is it necessary? Is it potentially*

capable of real predictive value? What are the limitations of current approaches, if any? (E.g.—is the fact that it is based on Reynolds averaging a fatal flaw? Will it always be necessary to have zonal models?) Is it essential to include, I or example, coherent structures explicitly in modeling?

7. *What role can we expect cellular automata to play in the future of turbulence? The accuracy and cost relative to finite difference schemes appear to be respectively low and high, but are there loopholes? Are current schemes Galilean invariant, and does it matter? Is there a niche where this approach can be useful?*

8. *The physics community appears to feel that most work on turbulence until the recent past was essentially engineering, and did not shed light on the heart of the problem. Many research managers in Washington seem to feel that, if this work was engineering, it was not responsive to program needs; only new and different approaches seem to attract support now. Are these perceptions of past accomplishments accurate, or are they a mix of specialty chauvinism, a failure to appreciate the difficulty of the turbulence problem, and the natural desire of a manager to make changes that will bear fruit during his tour of duty?*

• *The wake behind the cylinder exhibits a flow field which is chaotic in both space and time, but whose averages and statistical properties are stable. It is these averages and statistical properties that we want to understand.* Nelkin 1989. What Do We Know about Self-Similarity in Fluid Turbulence, J. Stat. Phys., **54**, 1–15.

• *We are not interested in details of the mechanism of energy input at small wave numbers (are they not important??). This problem has been nicely solved by Edwards* Nelkin, M. 1974 Turbulence, critical fluctuations, and intermittency *Phys. Rev.*, **A9**, 388–395.

• *What are the measures describing turbulence? The question arises because invariance under time evolution (the Hopf equation) is not enough to specify a unique probability measure which would describe turbulence. One knows indeed that strange attractors typically carry uncountably many disjoint invariant probability measures. The ones one wants probably satisfy stability under small stochastic perturbations. This is a powerful restriction, but hard to exploit,* Ruelle, D. 1984 Conceptual Problems of Weak and Strong Turbulence, Phys. Rep., **103**, 81–85.

• *However, it might well be possible to get a very good approximation to the average velocity curve with much less work than that required to take and average all the measurements. Even more importantly, a successful approximation method would almost certainly illuminate the physics of turbulent flow. These are the hopes on which turbulence closure theory rests.* Salmon, R. 1998 Lectures on geophysical fluid dynamics, Oxford University Press.

• *Do the Navier–Stokes equations on a 3-dimensional domain Ω have a unique smooth solution for all time?... The solution of this problem might well be a fundamental step toward the very big problem of understanding turbulence.* Smale, S., 1998 Mathematical problems for the next century, *The Mathematical Intelligencer*, **20**, 7–15.

References

Abe H, Antonia RA (2011) Scaling of normalized mean energy and scalar dissipation rates in a turbulent channel flow. Phys Fluids 23:055104

Aluie H (2012) Scale locality and the inertial range in compressible turbulence. J Fluid Mech (submitted). arXiv:1101.0150

Anderson PW (1972) More is different. Science 177:393–396

Armi L, Flament P (1987) Cautionary remarks on the spectral interpretation of turbulent flows. J Geophys Res 90:11,779–11,782

Arnold VI (1991) Kolmogorov's hydrodynamics attractors. Proc R Soc Lond A 434:19–22

Baig MF, Chernyshenko SI (2005) The mechanism of streak formation in near-wall turbulence. J Fluid Mech 544:99–131

Barbu V (2011) Stabilization of Navier–Stokes flows. Springer, Berlin

Bardos C, Titi ES (2007) Euler equations for incompressible ideal fluids. Russ Math Surv 62:409–451

Batchelor GK (1947) Kolmogoroff's theory of locally isotropic turbulence. Proc Camb Philol Soc 43:533–559

Batchelor GK (1953) The theory of homogeneous turbulence. Cambridge University Press, Cambridge

Batchelor GK (1962) The dynamics of homogeneous turbulence: introductory remarks. In: Favre A (ed) Mécanique de la turbulence, Marseille, 28 août–2 septembre 1961. Colloques internationaux du CNRS, vol 108, p 96

Batchelor GK (1989) Review: fluid mechanics. J Fluid Mech 205:593. By LD Landau and EM Lifshits, 2nd English edition

Batchelor GK, Townsend AA (1949) The nature of turbulent motion at large wave-numbers. Proc R Soc Lond A 199:238–255

Berdichevsky V, Fridlyand A, Sutyrin V (1996) Prediction of turbulent velocity profile in Couette and Poiseuille flows from first principles. Phys Rev Lett 76:3967–3970

Betchov R (1974) Non-Gaussian and irreversible events in isotropic turbulence. Phys Fluids 17:1509–1512

Betchov R (1976) On the non-Gaussian aspects of turbulence. Arch Mech 28(5–6):837–845

Betchov R (1993) In: Dracos T, Tsinober A (eds) New approaches and turbulence. Birkhäuser, Basel, p 155

Bevilaqua PM, Lykoudis PS (1978) Turbulence memory in self-preserving wakes. J Fluid Mech 89:589–606

Biferale L, Procaccia I (2005) Anisotropy in turbulent flows and in turbulent transport. Phys Rep 414:43–164

Biferale L, Lanotte AS, Federico Toschi F (2004) Effects of forcing in three-dimensional turbulent flows. Phys Rev Lett 92:094503

A. Tsinober, *The Essence of Turbulence as a Physical Phenomenon*,
DOI 10.1007/978-94-007-7180-2,

Blackwelder RF (1983) Analogies between transitional and turbulent boundary layers. Phys Fluids 26:2807–2815
Bonnet JP (ed) (1996) Eddy structure identification. Springer, Berlin
Borel E (1909) Sur les probabilites denombrables et leurs applications arithmetiques. Rend Circ Mat Palermo 41:247–271
Borisenkov Y, Kholmyansky M, Krylov S, Liberzon A, Tsinober A (2011) Super-miniature multi-hot-film probe for sub-Kolmogorov resolution in high-returbulence. J Phys Conf Ser 318:072004
Borodulin VI, Kachanov YS, Roschektayev AP (2011) Experimental detection of deterministic turbulence. J Turbul 12(23):1–34
Bradshaw P (1994) Turbulence: the chief outstanding difficulty of our subject. Exp Fluids 16:203–216
Cantwell BJ (1992) Exact solution of a restricted Euler equation for the velocity gradient tensor. Phys Fluids A 4:782–793
Chen Q, Chen S, Eyink GL, Holm DD (2003) Intermittency in the joint cascade of energy and helicity. Phys Rev Lett 90:214503
Chorin AJ (1994) Vorticity and turbulence. Springer, Berlin
Chorin AJ (1996) Turbulence cascades across equilibrium spectra. Phys Rev 54:2616–2619
Compte-Bellot G (1965) Ecoulement turbulent entre deux parois paralleles. In: Paris: publications scientifiques et techniques du ministere de l'air, vol 419, p 159. English translation: Bradshaw P (1969) In: Turbulent flow between two parallel walls. ARC no 31609. There is also a Russian translation
Constantin P (1996) Navier–Stokes equations and incompressible fluid turbulence. Lect Appl Math 31:219–234
Corrsin S (1958) Local anisotropy in turbulent shear flow. Natl Adv Com Aeronaut, Res Memo 58B11:1–15
Cullen MJP (2006) A mathematical theory of large-scale atmospheric flow. Imperial College Press, London
Cvitanović P, Gibson P (2010) Phys Scr T 142:014007
Doering CR (2009) The 3D Navier–Stokes problem. Annu Rev Fluid Mech 41:109–128
Doering CR, Gibbon JD (2004) Applied analysis of the Navier–Stokes equations. Cambridge University Press, Cambridge
Dowker M, Ohkitani K (2012) Intermittency and local Reynolds number in Navier–Stokes turbulence: a cross-over scale in the Caffarelli-Kohn-Nirenberg integral. Phys Fluids 24:115112
Dryden H (1948) Recent advances in boundary layer flow. Adv Appl Mech 1:1–40
Dwoyer DL, Hussaini MY, Voigt RG (eds) (1985) Theoretical approaches to turbulence. Springer, Berlin
Einstein A (1926) Letter to Max Born (4 December 1926); the Born-Einstein letters (translated by Irene Born). Walker and Company, New York. ISBN 0-8027-0326-7
Elliott FW, Majda AJ (1995) A new algorithm with plane waves and wavelets for random velocity fields with many spatial scales. J Comput Phys 117:146–162
Elsinga GE, Marusic I (2010) Universal aspects of small-scale motions in turbulence. J Fluid Mech 662:514–539
Eyink GL, Aluie H (2009a) Localness of energy cascade in hydrodynamic turbulence, I smooth coarse graining. Phys Fluids 21:115107
Eyink GL, Aluie H (2009b) Localness of energy cascade in hydrodynamic turbulence, II: sharp spectral filter. Phys Fluids 21:115108
Eyink GL, Frisch U (2011) Robert H. Kraichnan. In: Davidson PA, Kaneda Y, Moffatt K, Sreenivasan K (eds) A voyage through turbulence. Cambridge University Press, Cambridge, pp 329–372
Falkovich G (2009) Symmetries of the turbulent state. J Phys A, Math Theor 42:123001
Falkovich G, Xu H, Pumir A, Bodenschatz B, Biferale L, Boffetta G, Lanotte AS, Toschi F (2012) On Lagrangian single-particle statistics. Phys Fluids 24:055102

Ferchichi M, Tavoularis S (2000) Reynolds number dependence of the fine structure of uniformly sheared turbulence. Phys Fluids 12:2942–2953
Feynmann R (1963) Lect Phys 2:41–42
Foiaş C (1997) What do the Navier–Stokes equations tell us about turbulence? Contemp Math 208:151–180
Foiaş C, Manley O, Rosa R, Temam R (2001) Navier–Stokes equations and turbulence. Cambridge University Press, Cambridge
Frenkiel FN, Klebanoff PS, Huang TT (1979) Grid turbulence in air and water. Phys Fluids 22:1606–1617
Friedlander S, Pavlović N (2004) Remarks concerning modified Navier–Stokes equations. Discrete Contin Dyn Syst 10:269–288
Frisch U (1995) Turbulence: the legacy of A.N. Kolmogorov. Cambridge University Press, Cambridge
Frisch U, Orszag SA (1990) Turbulence: challenges for theory and experiment. Phys Today 43:24–32
Frisch U et al (2008) Hyperviscosity, Galerkin truncation, and bottlenecks in turbulence. Phys Rev Lett 101:144501
Gad-el-Hak M, Tsai HM (2006) Transition and turbulence control. World Scientific, Singapore
Galanti B, Tsinober A (2004) Is turbulence ergodic? Phys Lett A 330:173–180
Galanti B, Tsinober A (2006) Physical space helicity properties in quasi-homogeneous forced turbulence. Phys Lett A 352:141–149
George WK (2012) Asymptotic effect of initial and upstream conditions on turbulence. J Fluids Eng 134:061203
Germano M (1999) Basic issues of turbulence modeling. In: Gyr A, Kinzelbach W, Tsinober A (eds) Fundamental problematic issues in turbulence. Birkhäuser, Basel, pp 213–219
Germano M (2012) The simplest decomposition of a turbulent field. Physica D 241:284–287
Gibson CH, Stegen GS, Williams RB (1970) Statistics of the fine structure of turbulent velocity and temperature fields measured at high Reynolds numbers. J Fluid Mech 41:153–167
Gibson CH, Friehe CA, McConnell SO (1977) Structure of sheared turbulent fields. Phys Fluids 20(II):S156–S167
Gkioulekas E (2007) On the elimination of the sweeping interactions from theories of hydrodynamic turbulence. Physica D 226:151–172
Goldshtik MA, Shtern VN (1981) Structural turbulence theory. Dokl Akad Nauk SSSR 257(6):1319–1322 (in Russian)
Goldstein S (1969) ARFM 1:23
Goldstein S (1972) The Navier–Stokes equations and the bulk viscosity of simple gases. J Math Phys Sci (Madras) 6:225–261
Goto T, Kraichnan RH (2004) Turbulence and Tsallis statistics. Physica D 193:231–244
Grant HL, Stewart RW, Moilliet A (1962) Turbulence spectra from a tidal channel. J Fluid Mech 12:241–268
Grossman S (1995) Asymptotic dissipation rate in turbulence. Phys Rev E 51:6275–6277
Guckenheimer J (1986) Strange attractors in fluids: another view. Annu Rev Fluid Mech 18:15–31
Gulitskii G, Kholmyansky M, Kinzlebach W, Lüthi B, Tsinober A, Yorish S (2007a) Velocity and temperature derivatives in high Reynolds number turbulent flows in the atmospheric surface layer. Facilities, methods and some general results. J Fluid Mech 589:57–81
Gulitskii G, Kholmyansky M, Kinzlebach W, Lüthi B, Tsinober A, Yorish S (2007b) Velocity and temperature derivatives in high Reynolds number turbulent flows in the atmospheric surface layer. Part 2. Accelerations and related matters. J Fluid Mech 589:83–102
Gulitskii G, Kholmyansky M, Kinzlebach W, Lüthi B, Tsinober A, Yorish S (2007c) Velocity and temperature derivatives in high Reynolds number turbulent flows in the atmospheric surface layer. Part 3. Temperature and joint statistics of temperature and velocity derivatives. J Fluid Mech 589:103–123
Gylfason A, Ayyalasomayajula S, Warhaft Z (2004) Intermittency, pressure and acceleration statistics from hot-wire measurements in wind-tunnel turbulence. J Fluid Mech 501:213–229

Hamlington PE, Schumacher J, Dahm W (2008) Direct assessment of vorticity alignment with local and nonlocal strain rates in turbulent flows. Phys Fluids 20:111703

Hamlington PE, Krasnov D, Boeck T, Schumacher J (2012) Local dissipation scales and energy dissipation-rate moments in channel flow. J Fluid Mech 701:419–429

Hawking S, Penrose R (1996) The nature and time. Princeton University Press, Princeton, p 26

Hill RJ (1997) Applicability of Kolmogorov's and Monin's equations to turbulence. J Fluid Mech 353:67–81

Hill RJ (2002) Scaling of acceleration in locally isotropic turbulence. J Fluid Mech 452:361–370

Hill RJ (2006) Opportunities for use of exact statistical equations. J Turbul 7(43):1–13

Holmes PJ, Berkooz G, Lumley JL (1996) Turbulence, coherent structures, dynamical systems and symmetry. Cambridge University Press, Cambridge

Holmes PJ, Lumley JL, Berkooz G, Mattingly JC, Wittenberg RW (1997) Low-dimensional models of coherent structures in turbulence. Phys Rep 287:337–384

Hopf E (1948) A mathematical example displaying features of turbulence. Commun Pure Appl Math 1:303–322

Hopf E (1952) Statistical hydromechanics and functional calculus. J Ration Mech Anal 1:87–123

Hosokawa I (2007) A paradox concerning the refined similarity hypothesis of Kolmogorov for isotropic turbulence. Prog Theor Phys 118:169–173

Hou TY, Hua X, Hussain F (2013) Multiscale modeling of incompressible turbulent flows. J Comput Phys 232:383–396

Hoyle F (1957) The black cloud. Harper, New York

Hunt JCR, Carruthers DJ (1990) Rapid distortion theory and 'problems' of turbulence. J Fluid Mech 212:497–532

Hunt JCR, Eames I, Westerweel J, Davidson PA, Voropayev SI, Fernando J, Braza M (2010) Thin shear layers—the key to turbulence structure? J Hydro-Environ Res 4:75–82

Ishihara T, Hunt JCR, Kaneda Y (2011) Conditional analysis near strong shear layers in DNS of isotropic turbulence at high reynolds number. J Phys Conf Ser 318(4):042004

Jimenez J (2012) Cascades in wall-bounded turbulence. Annu Rev Fluid Mech 44:27–45

Kadanoff LP (1986) Fractals: where is the physics? Phys Today 39:3–7

Kawahara G, Uhlmann M, van Veen L (2012) The significance of simple invariant solutions in turbulent flows. Annu Rev Fluid Mech 44:203–225

Keefe L (1990a) Connecting coherent structures and strange attractors. In: Kline SJ, Afgan HN (eds) Near wall turbulence—1988 Zaric memorial conference. Hemisphere, Washington, pp 63–80

Keefe L (1990b) In: Lumley JL (ed) Whither turbulence? Springer, Berlin, p 189

Keefe L, Moin P, Kim J (1992) The dimension of attractors underlying periodic turbulent Poiseulle flow. J Fluid Mech 242:1–29

Keller L, Friedmann A (1925) Differentialgleichung für die turbulente Bewegung einer kompressiblen Flüssigkeit. In: Biezeno CB, Burgers JM (eds) Proceedings of the first international congress on applied mechanics. Waltman, Delft, pp 395–405

Kevlahan NK-R, Hunt JCR (1997) Nonlinear interactions in turbulence with strong irrotational straining. J Fluid Mech 337:333–364

Kholmyansky M, Tsinober A (2008) Kolmogorov 4/5 law, nonlocality, and sweeping decorrelation hypothesis. Phys Fluids 20:041704

Kholmyansky M, Tsinober A (2009) On an alternative explanation of anomalous scaling and how well-defined is the concept of inertial range. Phys Lett A 273:2364–2367

Kholmyansky M, Shapiro-Orot M, Tsinober A (2001c) Experimental observations of spontaneous breaking of reflexional symmetry in turbulent flows. Proc R Soc Lond 457:2699–2717

Kim J (2012) Progress in pipe and channel flow turbulence, 1961–2011. J Turbul 13(45):N45

Klewicki JC (2010) Reynolds number dependence, scaling and dynamics of turbulent boundary layers. J Fluids Eng 132:094001

Kolmogorov AN (1933) Grundbegriffe der Wahrscheinlichkeitsrechnung. Springer, Berlin. English translation: Kolmogorov AN (1956) Foundations of the theory of probability, Chelsea

Kolmogorov AN (1941a) The local structure of turbulence in incompressible viscous fluid for very large Reynolds numbers. Dokl Akad Nauk SSSR 30:299–303. For English translation see Tikhomirov VM (ed) (1991) Selected works of AN Kolmogorov, vol I, Kluwer, pp 318–321
Kolmogorov AN (1941b) Dissipation of energy in locally isotropic turbulence. Dokl Akad Nauk SSSR 32:19–21. For English translation see Tikhomirov VM (ed) (1991) Selected works of AN Kolmogorov, vol I, Kluwer, pp 324–327
Kolmogorov AN (1956) The theory of probability. In: Aleksandrov AD et al (eds) Mathematics, its content, methods and meaning. AN SSSR, Moscow. English translation: Am Math Soc, pp 229–264 (1963)
Kolmogorov AN (1962) A refinement of previous hypotheses concerning the local structure of turbulence is a viscous incompressible fluid at high Reynolds number. J Fluid Mech 13:82–85
Kolmogorov AN (1985) In: Notes preceding the papers on turbulence in the first volume of his selected papers, vol I. Kluwer, Dordrecht, pp 487–488. English translation: Tikhomirov VM (ed) (1991) Selected works of AN Kolmogorov
Kosmann-Schwarzbach Y, Tamizhmani KM, Grammaticos B (eds) (2004) Integrability of nonlinear systems. Lecture notes in physics, vol 638
Kraichnan (1957) The structure of isotropic turbulence at very high Reynolds numbers. J Fluid Mech 5:497–543
Kraichnan RH (1959) The structure of isotropic turbulence at very high Reynolds numbers. J Fluid Mech 5:497–543
Kraichnan RH (1961) Dynamics of nonlinear stochastic systems. J Math Phys 2(1):124–148
Kraichnan RH (1964) Kolmogorov's hypotheses and Eulerian turbulence theory. Phys Fluids 7:1723–1734
Kraichnan RH (1974) On Kolmogorov's inertial-range theories. J Fluid Mech 62:305–330
Kraichnan RH, Chen S (1989) Is there a statistical mechanics of turbulence? Physica D 37:160–172
Krogstad P-A, Antonia RA (1999) Surface effects in turbulent boundary layers. Exp Fluids 27:450–460
Kuo AY-S, Corrsin S (1971) Experiments on internal intermittency and fine-structure distribution functions in fully turbulent fluid. J Fluid Mech 50:285–319
Kush HA, Ottino JM (1992) Experiments in continuous chaotic flows. J Fluid Mech 236:319–348
Kuznetsov VR, Praskovsky AA, Sabelnikov VA (1992) Finescale turbulence structure of intermittent shear flows. J Fluid Mech 243:595–622
Ladyzhenskaya OA (1969) Mathematical problems of the dynamics of viscous incompressible fluids. Gordon and Breach, New York
Ladyzhenskaya OA (1975) Mathematical analysis of NSE for incompressible liquids. Annu Rev Fluid Mech 7:249–272
Lagrange J-L (1788) M'ecanique analitique, Paris, Sect. X, p 271
Lamb H (1932) Hydrodynamics. Cambridge University Press, Cambridge
Landau LD (1944) On the problem of turbulence. Dokl Akad Nauk SSSR 44:339–343 (in Russian). English translation in: Ter Haar D (ed) Collected papers of LD Landau, Pergamon, Oxford, pp 387–391
Landau LD (1960) Fundamental problems. In: Fierz M, Weisskopf VF (eds) Theoretical physics in the twentieth century, a memorial volume to Wolfgang Pauli. Interscience, New York, pp 245–247
Landau LD, Lifshits EM (1944) Fluid mechanics, 1st Russian edn
Landau LD, Lifshits EM (1959) Fluid mechanics. Pergamon, New York
Landau LD, Lifshits EM (1987) Fluid mechanics. Pergamon, New York
Laplace PS (1951) A philosophical essay on probabilities. Dover, New York. Translated by Truscott FW, Emory FL (Essai philosophique sur les probabilités. Rééd., Bourgeois, Paris, 1986. Texte de la 5éme éd., 1825)
Laval J-P, Dubrulle B, Nazarenko S (2001) Nonlocality and intemittency in three-dimensional turbulence. Phys Fluids 13:995–2012
Laws EM, Livesey JL (1978) Flow through screens. Annu Rev Fluid Mech 10:247–266

Leonov VP, Shiryaev AN (1960) Some problems in the spectral theory of higher order moments II. Theory Probab Appl 5:417–421
Leray J (1934) Essai sur le mouvement d'un fluide visqueux emplissant l'espace. Acta Math 63:193–248
Leung T, Swaminathan N, Davidson PA (2012) Geometry and interaction of structures in homogeneous isotropic turbulence. J Fluid Mech 710:453–481
Li Y, Perlman E, Wan M, Yang Y, Meneveau C, Burns R, Chen S, Szalay A, Eyink G (2008) A public turbulence database cluster and applications to study Lagrangian evolution of velocity increments in turbulence. J Turbul 9(31):1–29
Liepmann HW (1979) The rise and fall of ideas in turbulence. Am Sci 67:221–228
Liepmann HW (1997) A brief history of boundary layer structure research. In: Panton RL (ed) Self-sustaining mechanisms of wall turbulence. Comp Mech Publ, p 4
Lindborg E (1999) Can atmospheric kinetic energy spectrum be explained by two-dimensional turbulence? J Fluid Mech 388:259–288
Lions JL (1969) Quelques méthodes de résolution des problèmes uax limites non linéaires. Dunod Gauthier-Villars, Paris
Long RR (2003) Do tidal-channel turbulence measurements support k-5/3? Environ Fluid Mech 3:109–127
Lorenz EN (1963) Deterministic nonperiodic flow. J Atmos Sci 20:130–141
Lorenz EN (1972) Investigating the predictability of turbulent motion. In: Rosenblatt M, van Atta CC (eds) Statistical models and turbulence. Lecture notes in physics, vol 12, pp 195–204
Loskutov A (2010) Fascination of chaos. Phys Usp 53(12):1257–1280
Lu SS, Willmarth WW (1973) Measurements of the structure of the Reynolds stress in a turbulent boundary layer. J Fluid Mech 60:481–511
Lumley JL (1970) Stochastic tools in turbulence. Academic Press, New York
Lumley JL (1981) Coherent structures in turbulence. In: Meyer R (ed) Transition and turbulence. Academic Press, New York, pp 215–242
Lumley JL (1987) Review of 'Turbulence and random processes in fluid mechanics' by MT Landahl and E Mollo-Christensen. J Fluid Mech 183:566–567
Lumley JL (1989) The state of turbulence research. In: George WK, Arndt R (eds) Advances in turbulence. Hemisphere/Springer, Washington, pp 1–10
Lumley JL (1990) Future directions in turbulence research and the role of organized motion. In: Lumley JL (ed) Whither turbulence? Springer, Berlin, pp 97–131
Lumley JL, Yaglom AM (2001) A century of turbulence. Flow Turbul Combust 66:241–286
Lüthi B, Tsinober A, Kinzelbach W (2005) Lagrangian measurement of vorticity dynamics in turbulent flow. J Fluid Mech 528:87–118
Majda AJ, Kramer PR (1999) Simplified models for turbulent diffusion: theory, numerical modelling, and physical phenomena. Phys Rep 314:237–574
Malm J, Schlatter P, Sandham ND (2012) A vorticity stretching diagnostic for turbulent and transitional flows. Theor Comput Fluid Dyn 26:485–499
Martin PC (1976) Instabilities, oscillations, and chaos. J Phys I 1:C157–C166
Marusic I, McKeon BJ, Monkewitz PA, Nagib HM, Smits AJ, Sreenivasan KR (2010) Wall-bounded turbulent flows at high Reynolds numbers: recent advances and key issues. Phys Fluids 22:065103
McComb WD (1990) The physics of fluid turbulence. Clarendon, Oxford
McKeon BJ, Morrison JF (2007) Asymptotic scaling in turbulent pipe flow. Philos Trans R Soc Lond A 365(1852):635–876
Meneveau C (1991a) Analysis of turbulence in the orthonormal wavelet representation. J Fluid Mech 232:469–520
Meneveau C (1991b) Dual spectra and mixed energy cascade of turbulence in the wavelet representation. Phys Rev Lett 66:1450–1453
Meneveau C (2011) Lagrangian dynamics and models of the velocity gradient tensor in turbulent flows. Annu Rev Fluid Mech 43:219–245

Migdal AA (1995) Turbulence as statistics of vortex cells. In: Mineev VP (ed) The first Landau institute summer school, 1993. Gordon and Breach, New York, pp 178–204

Miles J (1984) Resonant motion of a spherical pendulum. Physica D 11:309–323

Moffatt HK (1990) In: Lumley JL (ed) Whither turbulence? Turbulence at the crossroads. Springer, Berlin, p 296

Mollo-Christensen E (1973) Intermittency in large-scale turbulent flows. Annu Rev Fluid Mech 5:101–118

Monin AS (1986) Hydrodynamic instability. Sov Phys Usp 29:843–868

Monin AS (1991) On definition of coherent structures. Sov Phys Dokl 36(6):424–426

Monin AS, Yaglom AM (1971) Statistical fluid mechanics, vol 1. MIT Press, Cambridge

Monin AS, Yaglom AM (1975) Statistical fluid mechanics, vol 2. MIT Press, Cambridge

Mullin T (ed) (1993) The nature of chaos. Clarendon, Oxford

Newton KA, Aref H (2003) Chaos vs turbulence. In: Scott A (ed) Encyclopedia of nonlinear science, pp 114–116

Novikov EA (1963) Random force method in turbulence theory. Zh Èksp Teor Fiz 44(6):2159–2168. English translation: Sov Phys JETP 17, 1449–1454 (1963)

Novikov EA (1967) Kinetic equations for a vortex field. Dokl Akad Nauk SSSR 177(2):299–301. English translation: Sov Phys Dokl 12(11), 1006–1008 (1968)

Novikov EA (1974) Statistical irreversibility of turbulence. Arch Mech 4:741–745

Novikov EA (1990a) The effects of intermittency on statistical characteristics of turbulence and scale similarity of breakdown coefficients. Phys Fluids A 2:814–820

Novikov EA (1990b) The internal dynamics of flows and formation of singularities. Fluid Dyn Res 6:79–89

Obukhov AN (1962) Some specific features of atmospheric turbulence. J Fluid Mech 13:77–81

Onsager L (1945) The distribution of energy in turbulence. Phys Rev 68:286

Onsager L (1949) Statistical hydrodynamics. Suppl Nuovo Cim VI(IX):279–287

Ornstein S, Weiss B (1991) Statistical properties of chaotic systems. Bull Am Math Soc 24:11–116

Orszag SA (1977) Lectures on the statistical theory of turbulence. In: Balian R, Peube J-L (eds) Fluid dynamics. Gordon and Breach, New York, pp 235–374

Orszag SA, Staroselsky I, Yakhot V (1993) Some basic challenges for large eddy simulation research. In: Orszag SA, Galperin B (eds) Large eddy simulation of complex engineering and geophysical flows. Cambridge University Press, Cambridge, pp 55–78

Ott E (1999) The role of Lagrangian chaos in the creation of multifractal measures. In: Gyr A, Kinzelbach W, Tsinober A (eds) Fundamental problematic issues in turbulence. Birkhäuser, Basel, pp 381–403

Palmer T (2005) Global warming in a nonlinear climate—can we be sure? Europhys News March/April:42–46

Palmer TN, Hardaker PJ (2011) Introduction: handling uncertainty in science. Philos Trans R Soc Lond A 369:4681–4684

Palmer T, Williams PD (2008) Introduction. Stochastic physics and climate modelling. Philos Trans R Soc Lond A 366:2421–2427

Pierrehumbert RT, Widnall SE (1982) The two- and three-dimensional instabilities of a spatially periodic shear layer. J Fluid Mech 114:59–82

Poincare H (1952a) Science and method. Dover, New York, p 286

Poincare H (1952b) Science and hypothesis. Dover, New York, pp xxiii–xiv

Pope SB (2000) Turbulent flows. Cambridge University Press, Cambridge

Porter DH, Woodward PR, Pouquet A (1998) Inertial range structures in decaying compressible turbulent flows. Phys Fluids 10:237–245

Pouransari Z, Speetjens MFM, Clercx HJH (2010) Formation of coherent structures by fluid inertia in three-dimensional laminar flows. J Fluid Mech 654:5–34

Priyadarshana P, Klewicki J, Treat S, Foss J (2007) Statistical structure of turbulent-boundary-layer velocity—vorticity products at high and low Reynolds numbers. J Fluid Mech 570:307–346

Pullin DI, Inoue M, Saito N (2013) On the asymptotic state of high Reynolds number, smooth-wall turbulent flows. Phys Fluids 25:015116

Pumir A, Shraiman BI, Siggia ED (1997) Perturbation theory for the δ-correlated model of passive scalar advection near the Batchelor limit. Phys Rev E 55:R1263
Richardson LF (1922) Weather prediction by numerical process. Cambridge University Press, Cambridge
Robinson JC (2001) Infinite-dimensional dynamical systems. Cambridge University Press, Cambridge
Robinson JC (2007) Parametrization of global attractors, experimental observations, and turbulence. J Fluid Mech 578:495–507
Rosteck AM, Oberlack M (2011) Lie algebra of the symmetries of the multi-point equations in statistical turbulence theory. J Nonlinear Math Phys 18(1):251–264
Ruelle D (1979) Microscopic fluctuations and turbulence. Phys Lett 72A(2):81–82
Ruelle D (1983) Differential dynamical systems and the problem of turbulence. Proc Symp Pure Math 39:141–154
Ruelle D (1984) Conceptual problems of weak and strong turbulence. Phys Rep 103:81–85
Ruelle D (1990) The turbulent fluid as a dynamical system. In: Sirovich L (ed) New perspectives in turbulence. Springer, Berlin, pp 123–138
Saddoughi SG (1997) Local isotropy in complex turbulent boundary layers at high Reynolds number. J Fluid Mech 348:201–245
Saffman PG (1968) Lectures on homogeneous turbulence. In: Zabusky NJ (ed) Topics in nonlinear physics. Springer, Berlin, pp 485–614
Saffman PG (1978) Problems and progress in the theory of turbulence. In: Fiedler H (ed) Structure and mechanics of turbulence, II. Lecture notes in physics, vol 76. Springer, Berlin, pp 274–306
Saffman PG (1991) In: Jimenez J (ed) The global geometry of turbulence. NATO ASI series B, vol 268. Plenum, New York, p 348
Salmon R (1998) Lectures on geophysical fluid dynamics. Oxford University Press, Oxford
Schumacher J, Kerr RM, Horiuti K (2011) Structure and dynamics of vorticity in turbulence. In: Davidson PA, Kaneda Y, Sreenivasan K (eds) Ten chapters in turbulence. Cambridge University Press, Cambridge, pp 40–79
Seiwert J, Morize C, Moisy F (2008) On the decrease of intermittency in decaying rotating turbulence. Phys Fluids 20:071702
She Z-S, Zhang Z-X (2009) Universal hierarchical symmetry for turbulence and general multi-scale fluctuation systems. Acta Mech Sin 25:279–294
She Z-S, Jackson E, Orszag SA (1990) Intermittent vortex structures in homogeneous isotropic turbulence. Nature 344:226–229
She Z-S, Chen X, Hussain F (2012) Lie-group derivation of a multi-layer mixing length formula for turbulent channel and pipe flows. arXiv:1112.6312v4 [physics.flu-dyn]
Shen X, Warhaft Z (2000) The anisotropy of the small-scale structure in high Reynolds number, $Re_\lambda = 1{,}000$, turbulent shear flow. Phys Fluids 12:2976–2989
Shlesinger MS (2000) Exploring phase space. Nature 405:135–137
Shraiman B, Siggia E (2000) Scalar turbulence. Nature 405:639–646
Shtilman L (1987) On one spectral property of the homogeneous turbulence. Unpublished manuscript
Shtilman L, Pelz R, Tsinober A (1988) Numerical investigation of helicity in turbulent flow. Comput Fluids 16:341–347
Siggia E (1994) High Rayleigh number convection. Annu Rev Fluid Mech 26:137–168
Sinai YaG (1999) Mathematical problems of turbulence. Physica A 263:565–566
Sirovich L (1997) Dynamics of coherent structures in wall bounded turbulence. In: Panton RL (ed) Self-sustaining mechanisms of wall turbulence. Comp mech publ, pp 333–364
Smits AJ, McKeon BJ, Marusic I (2011) High-Reynolds number wall turbulence. Annu Rev Fluid Mech 43:353–375
Southerland KB, Frederiksen RD, Dahm WJA (1994) Comparisons of mixing in chaotic and turbulent flows. Chaos Solitons Fractals 4(6):1057–1089
Sreenivasan KR, Antonia R (1997) The phenomenology of small-scale turbulence. Annu Rev Fluid Mech 29:435–472

Stewart RW (1969) Turbulence and waves in stratified atmosphere. Radio Sci 4:1269–1278
Suzuki Y, Nagano Y (1999) Modification of turbulent helical/nonhelical flows with small-scale energy input. Phys Fluids 11:3499–3511
Tan-Attichat J, Nagib HM, Loehrke RI (1989) Interaction of free-stream turbulence with screens and grids: a balance between turbulence scales. J Fluid Mech 114:501–528
Taylor GI (1935) The statistical theory of turbulence. Proc R Soc Lond A 151:421–478
Taylor GI (1938a) Production and dissipation of vorticity in a turbulent fluid. Proc R Soc Lond A 164:15–23
Taylor GI (1938b) The spectrum of turbulence. Proc R Soc Lond A 164:476–490
Tennekes H (1975) Eulerian and Lagrangian time microscales in isotropic turbulence. J Fluid Mech 67:561–567
Tennekes H (1976) Fourier-transform ambiguity in turbulence dynamics. J Atmos Sci 33:1660–1663
Tennekes H (1993) Karl Popper and the accountability of numerical weather forecasting. In: Workshop predictability and chaos in the geosciences, Boulder, 7–10 September 1993, pp 343–346
Tennekes H, Lumley JL (1972) A first course of turbulence. MIT Press, Cambridge
Tikhomirov VM (ed) (1991) Selected works of AN Kolmogorov, vol I. Kluwer, Dordrecht
Townsend AA (1948) Local isotropy in the turbulent wake of cylinder. Aust J Sci Res 1:161–174
Townsend AA (1976) The structure of turbulent shear flow. Cambridge University Press, Cambridge
Townsend AA (1987) Organized eddy structures in turbulent flows. Physicochem Hydrodyn 8(1):23–30
Tritton DJ (1988) Physical fluid dynamics, 2nd edn. Clarendon, Oxford
Truesdell C (1954) Kinematics of vorticity. Indiana University Press, Bloomington
Tsinober A (1995) Variability of anomalous transport exponents versus different physical situations in geophysical and laboratory turbulence. In: Schlesinger M, Zaslavsky G, Frisch U (eds) Levy flights and related topics in physics. Lecture notes in physics, vol 450. Springer, Berlin, pp 3–33
Tsinober A (1998a) Is concentrated vorticity that important? Eur J Mech B, Fluids 17:421–449
Tsinober A (1998b) Turbulence—beyond phenomenology. In: Benkadda S, Zaslavsky GM (eds) Chaos, kinetics and nonlinear dynamics in fluids and plasmas. Lecture notes in physics, vol 511. Springer, Berlin, pp 85–143
Tsinober A (2000) Vortex stretching versus production of strain/dissipation. In: Hunt JCR, Vassilicos JC (eds) Turbulence structure and vortex dynamics. Cambridge University Press, Cambridge, pp 164–191
Tsinober A (2001) An informal introduction to turbulence. Kluwer, Dordrecht
Tsinober A (2004) Helicity. In: Scott A (ed) Encyclopedia of nonlinear science, p 116
Tsinober A (2009) An informal conceptual introduction to turbulence. Springer, Berlin
Tsinober A, Galanti B (2003) Exploratory numerical experiments on the differences between genuine and 'passive' turbulence. Phys Fluids 15:3514–3531
Tsinober A, Vedula P, Yeung PK (2001) Random Taylor hypothesis and the behaviour of local and convective accelerations in isotropic turbulence. Phys Fluids 13:1974–1984
Van Zandt TE (1982) A universal spectrum of buoyancy waves in the atmosphere. Geophys Res Lett 9:575–578
Vassilicos JC (ed) (2001) Intermittency in turbulent flows. Cambridge University Press, Cambridge
Vedula P, Yeung PK (1999) Similarity scaling of acceleration and pressure statistics in numerical simulations of isotropic turbulence. Phys Fluids 11:1208–1220
Vishik MJ, Fursikov AV (1988) Mathematical problems of statistical hydromechanics. Kluwer, Dordrecht
von Karman T (1943) Tooling up mathematics for engineering. Q Appl Math 1(1):2–6
von Karman Th, Howarth L (1938) On the statistical theory of isotropic turbulence. Proc R Soc Lond Ser A, Math Phys Sci 164:192–215
von Neumann J (1949) Recent theories of turbulence. In: Taub AH (ed) A report to the office of naval research. Collected works, vol 6. Pergamon, New York, pp 437–472

Vukasinovich B, Rusak Z, Glezer A (2010) Dissipative small-scale actuation of a turbulent shear layer. J Fluid Mech 656:51–81

Waleffe F (1992) The nature of triad interactions in homogeneous turbulence. Phys Fluids A 4:350–363

Wei T, Willmarth WW (1989) Reynolds-number effects on the structure of a turbulent channel flow. J Fluid Mech 204:57–95

Wiener N (1938) Homogeneous chaos. Am J Math 60:897–936

Wolf M, Lüthi B, Holzner M, Krug D, Kinzelbach W, Tsinober A (2012a) Investigations on the local entrainment velocity in a turbulent jet. Phys Fluids 24:105110

Wolf M, Lüthi B, Holzner M, Krug D, Kinzelbach W, Tsinober A (2012b) Effects of mean shear on the local turbulent entrainment process. J Fluid Mech (in press)

Worth NA, Nickels TB (2011) Some characteristics of thin shear layers in homogeneous turbulent flow. Philos Trans R Soc Lond A 2011(369):709–722

Wygnanski I, Champagne F, Marasli B (1986) On the large scale structures in two-dimensional, small-deficit, turbulent wakes. J Fluid Mech 168:31–71

Yaglom AM (1949) On the field of accelerations in turbulent flow. Dokl Akad Nauk SSSR 67:795–798

Yakhot V, Orszag SA (1987a) Renormalization group and local order in strong turbulence. Nucl Phys B, Proc Suppl 2:417–440

Yakhot V et al (1987b) Weak interactions and local order in strong turbulence. Annu Rev Fluid Mech 20

Yudovich VI (2003) Eleven great problems of mathematical hydrodynamics. Mosc Math J 3:711–737

Zakharov VE (ed) (1990) What is integrability? Springer, Berlin

Zaslavsky GM (1999) Chaotic dynamics and the origin of statistical laws. Phys Today 51:39–45

Zeldovich YaB (1979) In: Barenblatt GI (ed) Similarity, self-similarity and intermediate asymptotics. Consultant Bureau, New York, p viii

Zeldovich YaB, Ruzmaikin AA, Sokoloff DD (1990) The almighty chance. World Scientific, Singapore

Author Index

A. Tsinober, *The Essence of Turbulence as a Physical Phenomenon*,
DOI 10.1007/978-94-007-7180-2,

Subject Index

A. Tsinober, *The Essence of Turbulence as a Physical Phenomenon*,
DOI 10.1007/978-94-007-7180-2,

Lightning Source UK Ltd.
Milton Keynes UK
UKOW07n2001220216

268865UK00001B/3/P

9 789400 771796